T0202991

SpringerBriefs in Molecular Science

SpringerBriefs in Molecular Science present concise summaries of cutting-edge research and practical applications across a wide spectrum of fields centered around chemistry. Featuring compact volumes of 50 to 125 pages, the series covers a range of content from professional to academic. Typical topics might include:

- A timely report of state-of-the-art analytical techniques
- A bridge between new research results, as published in journal articles, and a contextual literature review
- A snapshot of a hot or emerging topic
- An in-depth case study
- A presentation of core concepts that students must understand in order to make independent contributions

Briefs allow authors to present their ideas and readers to absorb them with minimal time investment. Briefs will be published as part of Springer's eBook collection, with millions of users worldwide. In addition, Briefs will be available for individual print and electronic purchase. Briefs are characterized by fast, global electronic dissemination, standard publishing contracts, easy-to-use manuscript preparation and formatting guidelines, and expedited production schedules. Both solicited and unsolicited manuscripts are considered for publication in this series.

More information about this series at http://www.springer.com/series/8898

Svatopluk Civiš · Martin Ferus · Antonín Knížek

The Chemistry of CO_2 and TiO_2

From Breathing Minerals to Life on Mars

 Springer

Svatopluk Civiš(iD)
Department of Spectroscopy
J. Heyrovský Institute of Physical
Chemistry, Czech Academy
of Sciences, v.v.i.
Prague, Czech Republic

Martin Ferus
Department of Spectroscopy
J. Heyrovský Institute of Physical
Chemistry, Czech Academy
of Sciences, v.v.i.
Prague, Czech Republic

Antonín Knížek(iD)
Department of Spectroscopy
J. Heyrovský Institute of Physical
Chemistry, Czech Academy
of Sciences, v.v.i.
Prague, Czech Republic

ISSN 2191-5407   ISSN 2191-5415 (electronic)
SpringerBriefs in Molecular Science
ISBN 978-3-030-24031-8   ISBN 978-3-030-24032-5 (eBook)
https://doi.org/10.1007/978-3-030-24032-5

This Springer imprint is published by the registered company Springer Nature Switzerland AG
The registered company address is: Gewerbestrasse 11, 6330 Cham, Switzerland

Someone else always has to carry on the story.
J. R. R. Tolkien

To all pilgrims on the path of knowledge…

Preface

A broad range of experimental and theoretical work has been devoted to the interaction of carbon dioxide with titania (TiO_2) in recent years, not least because of the changing state of the global climate. Carbon dioxide chemistry and TiO_2 surface catalysis nowadays find their use in many applications from battling the global warming through technical application, fuel production and energy storage to origin of life studies and synthesis of organic molecules on early Earth, Mars and other terrestrial planets.

The purpose of this book is to show two important aspects of the interaction between gaseous CO_2 and the semiconductor mineral surface. Firstly, this book will show that oxygen atoms are exchanged between the oxygen-containing minerals and CO_2 in ambient conditions. This surprising feature which is often neglected, shows that the boundary between the gas phase and the solid phase is not inert and that room temperature may be sufficient to allow the interaction to take place.

Furthermore, this oxygen atom exchange is not limited to titanium dioxide. Other minerals may interact with the gas in a similar manner, including many commonly occurring natural minerals. It seems, therefore, that this surprising feature is not unique. Quite the contrary, it is a commonplace process, which affects the ongoing chemistry in the environment and should be considered in every model that features natural processes.

Secondly, in the past four decades, many studies have shown that carbon dioxide can be readily reduced to methane through a process called 'methanogenesis.' The photocatalytic reduction of carbon dioxide to methane, methanol and other carbon-containing compounds is an important process for our planet. If employed on a large-scale basis, it may reduce carbon dioxide content in the atmosphere, capture solar energy, which is the ultimate inexhaustible source of energy (at least until humankind migrates outside the Solar system, if that will ever happen) and produce fuel or chemical reagents. This process is often coupled with water splitting, which produces hydrogen and oxygen, which in turn can again be used as fuels or for chemical synthesis.

The area of research into these solar-powered processes is very big and still growing exponentially. Most work published on this topic, however, is concerned with the material chemistry, doping, photocatalyst efficiency and sensitizing. This book, on the other hand, will show the chemical aspects of the methanogenesis and discuss how they could be exploited to enhance the effectivity of the process. For example, the effect of an acidic proton in the environment will be discussed and we will describe that acidic conditions boost the rate of the reduction reaction.

Such findings can serve a purpose in the environmental chemistry research indeed, but there is a staggering amount of literature on this topic already available. What this book will show is that natural minerals can be used as photocatalysts for the methanogenesis as well and that it may occur in the natural environment on contemporary Earth. Looking back in time, we shall see that this process may have, or indeed, must have taken place on the early Earth, where it must have influenced the redox state of the atmosphere. Consequently, organic synthesis or prebiotic synthesis could have taken place in ways unprecedented to this day and the chemistry of CO_2 must have played an important role in the chemical origin of life.

What is more, we will show that this process may be used for the explanation of the presence of methane on Mars and may shed some light on the seasonal variation of CH_4, which has been observed recently on Mars. The methanogenesis is probably insufficient to explain these phenomena completely, but it definitely plays its part and again, should be included in any models of planetary chemistries such as there may be.

Overall, this book should neither serve as a definitive answer to humankind problems and resolve the issue of the global change of climate nor should it be used as a manual for the designing of photoreactors. For this, there are books and articles more fitting than our work. Rather this book should be able to show some of the often neglected aspects and processes and review their potential role in the chemistry of the contemporary and early Earth and beyond.

Acknowledgements This work was partly funded by the Czech Science Foundation (grant No. 19-03314S). Part of this work was also financed by project GAUK 16742 and ERDF/ESF "Centre of Advanced Applied Sciences" (No. CZ.02.1.01/0.0/153 0.0/16_019/0000778).

Prague, Czech Republic   Svatopluk Civiš
                                                                                       Martin Ferus
                                                                                  Antonín Knížek

Contents

Chapter 1
Carbon Dioxide and the Effects on Climate

Abstract Last a few decades have witnessed a rising scientific interest in two well-known molecules—carbon dioxide and titania. The fundamental properties such as their chemistry, structure and application potential are well known. However, the main problem concerning CO_2—the global change of climate—has not been dealt with yet. The ever-rising levels of carbon dioxide in the atmosphere, utilization of fossil fuels and the effects on global climate portend the need for a cleaner energy supply as well as the need to reduce the CO_2 amount present in the atmosphere of our planet. Importantly, several recent discoveries have pointed out that novel materials based on TiO_2 are able to solve climate change by establishing a new methane cycle-based energetics and economy.

1.1 Harvesting Solar Energy

Much focus is nowadays directed towards the exploration of novel methods for harvesting energy from renewable sources. Among the most promising technologies is the conversion of solar energy into chemical energy in the form of 'fossil fuels'. As the Sun is an inexhaustible source of energy from the humankind's point of view, the utilization of solar power seems to be an ideal way of developing a sustainable energy infrastructure (Inoue et al. 1979).

There exist various possibilities of solar energy utilization such as thermal reactions and water splitting. One of the most popular is photocatalysis, where incident photons from the Sun get absorbed in the photocatalyst and an electron is excited from the valence band into the conduction band. In this way, the solar energy is converted into chemical energy (Pei and Luan 2012; Zhou et al. 2014; White et al. 2015; Chang et al. 2016; Sohn et al. 2017) which can then be harvested through powering various reactions, such as CO_2 photocatalytic reduction. The total solar energy reaching the Earth's surface in 1 h roughly equals the annual energy consumption by humans (Lewis and Nocera 2006). Harvesting this energy would therefore be ideal. Converting as little as 10% of the solar energy on 0.3% land area of our planet would be sufficient to fulfil (and indeed, exceed) the projected energy needs of the world's population in 2050.

© The Author(s), under exclusive license to Springer Nature Switzerland AG 2019
S. Civiš et al., *The Chemistry of CO2 and TiO2*,
SpringerBriefs in Molecular Science,
https://doi.org/10.1007/978-3-030-24032-5_1

1

Table 1.1 Global annual CO_2 emissions from 1960 till 2017

Year	Global CO_2 emissions (Gt)
1960	9.4
1970	14.79
1980	19.32
1990	22.15
2000	24.69
2009	31.89
2011	34.85
2013	35.84
2015	36.02
2017	36.79

This table was compiled from data available on (Hausfather 2017; Carbon Dioxide Information Analysis Center, Environmental Sciences Division, Oak Ridge National Laboratory, Tennessee 2019)

Current usage of fossil fuels for energy production resulted in the emission of 30.4 Gt of carbon dioxide into the atmosphere (Centi and Perathoner 2010) in 2010 and 35.9 Gt in 2014 (CO_2-earth). The emissions of CO_2 in the past 60 years are summarized in Table 1.1.

Total carbon dioxide emissions reached 36.7 Gt of carbon dioxide in 2017 (CO_2-earth), and this number is likely to increase to over 40 Gt by 2035, depending on the energy and land use implemented by governments across the world (Lewis and Nocera 2006). The natural carbon dioxide cycle involves about 90 Gt of CO_2, where both emission and fixation (plants, microorganisms, underground sequestration) take part (Aresta and Dibenedetto 2007). It is obvious that over 35 Gt of anthropogenic CO_2 emitted will disturb the balance. The atmospheric level rose from 270 ppm in the preindustrial era to over 412 ppm in 2019 (CO_2-earth). The observed concentration also far exceeds the natural fluctuation (180–300 ppm) monitored over the past 800,000 years (Smol 2012) and is probably the highest in the past 15 million years (Pearson and Palmer 2000). Since the rise in CO_2 levels is of anthropogenic cause, it is the humankind's responsibility to search for a solution to save the current state of the climate (Mikkelsen et al. 2010; Maginn 2010).

1.2   The CO_2 Molecule

CO_2 is a colourless gas, which constitutes of two oxygen atoms double bonded to a central carbon atom. The molecule is linear in the ground state, and the C=O bond length is 116.3 pm. Being centrosymmetric, the molecule has no electrical dipole. The molecule has 4 normal modes of oscillation, two of which are degenerate. The antisymmetric stretching mode can be observed at $2349 \, \text{cm}^{-1}$ in the infrared spectrum

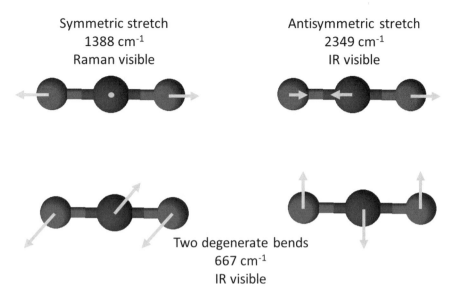

Symmetric stretch
1388 cm^{-1}
Raman visible

Antisymmetric stretch
2349 cm^{-1}
IR visible

Two degenerate bends
667 cm^{-1}
IR visible

Fig. 1.1 Vibrational modes of CO$_2$. Vibrational modes where the symmetry of the excited state is the same as the symmetry of the transition dipole moment operator are allowed transitions. If the transition moment operator transforms in x, y or z, the transition is visible by infrared spectroscopy. If the transformation of the transition moment operator is of second order (such as xy, xz and yz), the transition is observable in Raman spectroscopy. Simply put, transitions that are active and observable in infrared spectroscopy are not visible in Raman spectroscopy and vice versa

and the degenerate bending modes at 667 cm^{-1}. The symmetric stretching mode lies at 1388 cm^{-1} and is observed in Raman spectroscopy. The vibrational modes are shown in Fig. 1.1.

CO$_2$ is an important molecule in many different areas. In the atmosphere, where currently, there is about 412 ppm of CO$_2$, it is responsible for maintaining the greenhouse effect. This means that CO$_2$ traps radiation scattered from the surface and heats up the planet. Simply put, the more CO$_2$, the warmer the planet will become. Stronger evaporation of the oceans implies higher water content in the atmosphere. Since water traps IR radiation (water is a greenhouse gas, in principle), the climate enters a spiral towards higher and higher temperatures. Warmer climate also means stronger evaporation of oceans, oscillations in weather or the acidification of oceans. CO$_2$ is a weak electrophile, and in aqueous solution, CO$_2$ forms carbonic acid (H$_2$CO$_3$). The carbonic acid is sometimes called 'respiratory acid', because it is the only acid excreted by human breath as a gas. It is a weak acid and its creation is not fast, which means that in a solution of CO$_2$ and H$_2$O, most carbon remains as CO$_2$. However, the higher the CO$_2$ pressure, the more CO$_2$ is dissolved into water, which in turn means more H$_2$CO$_3$. This is the primary cause of the acidification of oceans and the massive extinction of corals. It should be noted that the change in ocean pH is currently estimated to be 0.1 on the pH scale relative to the preindustrial

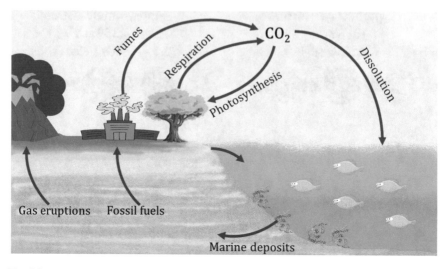

Fig. 1.2 A simplified carbon dioxide cycle on Earth

era and the ocean remains basic. Still, the change is significant for the fauna and flora of the ocean.

CO_2 and water were the most important atmospheric constituents after the primary hydrogen-helium atmosphere from the protoplanetary nebula disappeared. Most of the carbon on Earth is nowadays deposited in carbonates in the Earth's mantle. If all this carbon were to be liberated as CO_2, it would give an atmosphere of 210 bars of CO_2. The annual natural cycle encompasses 90 Gt of CO_2, so if humankind produces over 35 Gt of CO_2, a disturbance in the cycle is inevitable. A schematic carbon dioxide cycle is shown in Fig. 1.2.

CO_2 is a terminal carbon oxidation product and is created by the combustion of more reduced carbon, most importantly methane and by the calcination of $CaCO_3$. Photosynthesizing organisms utilize atmospheric CO_2 and convert it through photosynthesis to complex molecules, such as glucose. The side product of this process is molecular oxygen. Deforestation, therefore, lowers the nature's capacity to trap CO_2 and reduce its atmospheric levels. Most CO_2 is utilized by oceanic phytoplankton and by tress (e.g. in the Amazonian rainforest).

Natural sinks of CO_2 are not enough to cover the human emissions of this atmospheric gas. Utilization and storage of energy in CO_2 through its reduction to methane would be an ideal solution to the energetic demands and the atmospheric changes that we now face.

Fig. 1.3 A hydrogen-powered bicycle. The bicycle is manufactured by Pearl (SPHPST.Co) and was first produced in 2007. The image was made by Pearl Hydrogen [CC BY-SA 3.0 (https://creativecommons.org/licenses/by-sa/3.0)]

1.3   Energy Storage and Beyond

The storage of energy in the form of hydrogen (gaseous or liquid) is an effective way of energy storage in a sustainable way. Molecular hydrogen obtained from water splitting using sunlight is presumably the cleanest way towards the development of low carbon emission economy. The hydrogen molecule contains a high amount of energy relative to its mass, and at the same time, its burning produces water and no other greenhouse gases. The research in hydrogen economy has been carried out for the past 47 years, ever since Fujishima and Honda first reported water splitting on a TiO_2 electrode (Fujishima and Honda 1972). Hydrogen as a fuel is nowadays used in the car industry, public transport such as buses, ships and also bicycles (Fig. 1.3).

Another promising technique is the storage of solar energy through photochemical reduction of CO_2 in high-energy compounds, such as methane, CO or methanol. Such 'artificial photosynthesis' utilizing solar fuels is a carbon neutral process (no additional carbon used) (Ma et al. 2014). CH_4 can then be used as a fuel or for the synthesis of acetylene. CO is used as a meat colouring agent or phosgene synthesis reagent, and methanol is used as a reagent for organic synthesis, a solvent or a fuel.

The current research into the photocatalytic reduction of CO_2 is far from complete, and there remain significant breakthroughs to be made. This technique, in order to be economically efficient and to have a positive energy balance, has to be reliant exclusively on solar radiation. The major issue with this approach is the fact that the vast majority of the photocatalysts that show efficiency in the photocatalytic reduction do not exhibit significant response to visible light. The reason for this is the need for a band gap sufficiently large to activate CO_2. In fact, most reviews published up to this day discuss titania, usually in anatase or rutile phases (Dhakshinamoorthy et al. 2012; Habisreutinger et al. 2013; Ma et al. 2014; Li et al. 2016; Abdullah et al. 2017; Shehzad et al. 2018), which has a band gap of approximately 3.0–3.2 eV. Being one of the most studied photocatalysts, TiO_2 is already widely used in environmental purification (Frank and Bard 1977), artificial photosynthesis, hydrogen production, self-cleaning, organic synthesis, solar cells (Li et al. 2014, 2016), etc. It is a relatively cheap, non-toxic material and as such is an ideal model for photocatalysis-oriented research. Consequently, the number of publications on the properties and use of TiO_2 has surged in the past decades. Not only are there reviews focused mainly on TiO_2, its doping, or development as a photocatalyst [such as Habisreutinger et al. (2013)], but there are also reviews on photocatalysts, metal oxides and semiconductors in general, which still contain vast descriptions of TiO_2 photocatalytic properties (Hoffmann et al. 1995; Matsuoka et al. 2007).

It is obvious that TiO_2 is widely applied across various fields of research and technology. Nevertheless, the most promising and yet challenging application is the photocatalytic reduction of CO_2 to produce fuels. This book will therefore focus on the recent research progress in the study of the interaction of CO_2 with TiO_2 surfaces and the mobility of CO_2 and oxygen between the gas phase and the photocatalyst surface and structure. Different to the vast majority of the published work, that is often concerned with the technological improvement of the photocatalyst itself, we attempt to describe the chemical aspects of the process which may help make the process of the photocatalytic reduction of CO_2 more effective.

Following the first chapters concerned with the theoretical and experimental explanations of the fundamental processes on the TiO_2 surface, we will focus on experimental applications. We will explore the oxygen mobility on different crystal structures and different minerals. A variety of natural minerals was used, and the same behaviour as in TiO_2 was proven, sometimes even more efficient. This shows that the boundary between the solid and gas phases is a dynamical site with interaction taking place even at ambient conditions.

Next, we will describe the photocatalytic reduction of CO_2 to methane and CO, again using a variety of mineral catalysts. Aside from the experimental description and the description of physico-chemical properties, we will explore the application of this process to the chemistry of Earth and Mars and will show that the methanogenesis is closely bound with the origin of perchlorates on Mars.

Latter chapters of this book will show the photocatalytic reduction of CO_2 to methane in the context of planetary atmospheres and their possible reprocessing by high energy density events, such as asteroid impacts, whereupon complex organic molecules are created. The discussion of the nanotechnology, photochemistry and surface processes in the context of the recent research on planets, for example on Mars (Curiosity rover by NASA), introduces a fundamental connection between the chemistry of the microscopic and the macroscopic, even extraterrestrial, worlds.

A follow-up to this connection is the application of these results to the origin of life research. In several studies, we attempted to join the experimental results of the photocatalytic reduction of CO_2 and the contemporary knowledge of the possible origin of life on Earth to introduce a new scenario for the production of biologically relevant molecules, namely nucleic acid bases. The results will be shown here and will add context to the fundamental research discussed in the first chapters.

Chapter 2
Oxygen Atoms Exchange Between Carbon Dioxide and TiO$_2$ (Light Induced and Spontaneous)

Abstract One of the exceptional features of CO$_2$ is its ability to interact with solid surfaces on a common basis. This interaction takes the form of an adsorption and can be monitored by isotope labelling. Oxygen isotope-labelled CO$_2$ added to a mineral, most importantly TiO$_2$, exchanges its oxygen atoms with the sample and the changing isotopic composition can be monitored by high-resolution Fourier transform infrared spectroscopy.

2.1 The Isotope Labelling of Minerals and CO$_2$

The necessary basics for the understanding of the photocatalytic CO$_2$ reduction on TiO$_2$ require a thorough knowledge of the interaction of CO$_2$ and the mineral surface. A lot of work has already been done on this topic (Thompson et al. 2004). As it turns out, isotope labelling provides a convenient tool for the study of the surface chemistry. The foundations of isotopic labelling of TiO$_2$ for the purpose of the observation of the surface chemistry were laid in the middle of the twentieth century (Sato 1987; Yanagisawa and Ota 1991; Yanagisawa and Sumimoto 1994; Yanagisawa 1995; Henderson 1996; Muggli and Falconer 1999; Liao et al. 2002; Henderson et al. 2003; Wu et al. 2008). The original method uses the most common isotope, Ti^{16}O$_2$, and puts it in contact with gaseous reactants such as H$_2$^{18}O, ^{18}O$_2$ and their ^{16}O-counterparts. The isotope exchange reaction then involves a replacement of the lattice ^{16}O with ^{18}O from the reactant. Especially, the reactions of isotopically labelled water allowed for the probing of common catalytic reactions (Pichat et al. 2007; Felipe Montoya et al. 2011; Mikhaylov et al. 2012). The same experimental approach is also applicable to molecules such as formic acid (Civiš et al. 2011; Kavan et al. 2011), alcohols (Civiš et al. 2012), carbon monoxide and carbon dioxide (Yanagisawa and Sumimoto 1994; Hebenstreit et al. 2000; He et al. 2009; Civiš et al. 2011) or carbonates (Civiš et al. 2011), where isotopically labelled oxygen enables revelations concerning the oxygen mobility and the reaction mechanism.

A different approach to employing isotopically labelled gaseous reactants is the use of isotopically labelled Ti^{18}O$_2$ or 'classical' reactants, such as ^{12}C^{18,18}O$_2$ or ^{13}C^{16,16}O$_2$. The synthesis of this material is straightforward as it involves only expo-

S. Civiš et al., *The Chemistry of CO$_2$ and TiO$_2$*,
SpringerBriefs in Molecular Science,
https://doi.org/10.1007/978-3-030-24032-5_2

sure of the non-labelled mineral to either $^{18}O_2$ or $H_2^{18}O$ at 750 K (Cheng and Selloni 2009) and a simultaneous UV irradiation (Civiš et al. 2012). An alternative synthetic way is the electrochemical oxidation in an ^{18}O-enriched electrolyte (Kavan et al. 2011). This process is limited to the surface layers of the mineral. High temperature creates oxygen vacancies and undercoordinated binding sites for the reactant, and UV irradiation provides sufficient energy for the excitation of an electron in the material (the band gap of TiO_2 is approx. 3.0–3.2 eV depending on the crystal structure). For instance, the reaction of $^{18}O_2$ with the $Ti^{16}O_2$ yielded $^{18}O/^{16}O$ ratio of 2.5 in the surface layer (Cheng and Selloni 2009). Reports of pure $Ti^{18}O_2$ are relatively rare.

The first step in our work, and the one that proved to be the most important, since it revealed a surprising feature of many natural compounds, was the study of oxygen mobility ($^{18}O/^{16}O$) between gaseous CO_2 and solid TiO_2. We have employed the isotopically labelled $C^{18,18}O_2$ together with $Ti^{16}O_2$ and also, for the first time, $C^{16,16}O_2$ and $Ti^{18}O_2$. It should be noted here, that carbon dioxide is a quite advantageous molecule for the isotope exchange studies and hence has been selected. Firstly, CO_2 has a rovibrational band at around 2349 cm^{-1} and two combination bands in 3500–3800 cm^{-1}. Using these highly resolved rotational–vibrational bands, it is easy to monitor even sub-ppm concentrations of the gas as well as monitor the isotopic composition of the gas due to a shift in the bands' maxima of each isotope. Secondly, CO_2 is a stable molecule and a final product of carbon oxidation (both organic and inorganic). The presented measurements had, as a primary goal, the determination of the rate constants of the room temperature spontaneous isotopic exchange between the gas and the mineral surface. With the use of the kinetic data as well as vibrational spectra analyses, a tentative mechanism of the exchange can be proposed as well.

2.2 Light-Induced Isotopic Exchange Between CO_2 and $Ti^{18}O_2$

The first attempts to study the light-induced oxygen exchange between CO_2 and TiO_2 were connected with the application of high-power excimer lasers, such as ArF and XeCl. The following results were originally published in Civiš et al. (2011). The synthesis of $Ti^{18}O_2$ was carried out in a closed all-glass vacuum apparatus. Titanium tetrachloride was twice distilled in the vacuum before use. One gram of $H_2^{18}O$ (^{18}O 97%) was cooled with liquid N_2 in high vacuum, and the ice was put into contact with 2.8 ml of $TiCl_4$ vapour using a glass valve. When both reactants were mixed, the mixture was taken out of the cooling bath and the reaction mixture was left to warm up at room temperature for 14 h. The produced HCl was siphoned to a side glass vial cooled to 77 K. Subsequently, the solid product was annealed overnight at 200 °C in a sealed vacuum apparatus. The HCl trap vial was kept at 77 K during the annealing. After the annealing, the HCl-containing vial was sealed off. Finally, the apparatus was unsealed in an Ar atmosphere in a glove box and the solid white powder was taken out. This sample is henceforth designated as T200. For storage

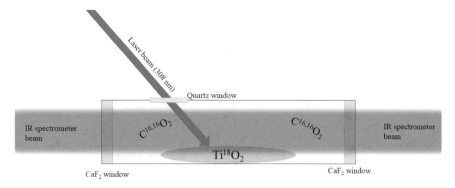

Fig. 2.1 Sample cuvette and its experimental set-up during the experiment with $Ti^{18}O_2$ and $C^{16,16}O_2$

purposes, the material was kept in a closed vial filled with Ar. Before the exchange experiment itself, a part of the T200 powder was annealed at 450 °C in 10^{-5} Pa vacuum for 30 h. Thus, modified sample was re-designated as T450. Both samples were characterized by X-ray diffraction (Bruker D8 Advance diffractometer; CuKα radiation) and exhibited the pattern of pure anatase. BET from surface areas were calculated from N_2 adsorption isotherms (Micromeritics ASAP 2020 instrument). The samples were degassed at 400 °C in the vacuum prior to measurement. The BET surface area was found to be 31 $m^2\ g^{-1}$ independent of the annealing history.

Light-induced isotopic exchange between $C^{16,16}O_2$ and $Ti^{18}O_2$ was studied using $Ti^{18}O_2$ annealed in a vacuum at 200 °C only (sample T200, Fig. 2.1). The 20 cm long optical cell was filled on a vacuum line with 0.8 g of the T200 powder and subsequently with 2 Torr of $C^{16,16}O_2$. The obtained spectrum in the region of 2000–4000 cm^{-1} is shown in Fig. 2.2, curve b) together with the reference spectrum of carbon dioxide in titania-free optical cell (Fig. 2.2, curve a). A rotational–vibrational band of HCl was detected in the 2800–3000 cm^{-1} spectral region, and small amount of water (mostly $H_2^{16}O$) is apparent by the band at 3600–3800 cm^{-1} (Fig. 2.2 curve b). Hydrogen chloride is obviously an impurity in the T200 sample: HCl is a product of the synthesis of T200 (reaction of $TiCl_4$ with $H_2^{18}O$), and by calcination at 200 °C in a closed apparatus, it is not quantitatively removed. The adsorbed HCl is released from the sample into the gas phase of our optical cell. The desorption of HCl from T200 occurs already in dark (Fig. 2.2, panel b) and progresses upon illumination with UV laser (Fig. 2.2, panel c–d).

However, upon UV photoexcitation, we observe additional processes beyond the HCl desorption. In the experiment, the T200 sample surface was irradiated with a XeCl laser (308 nm, 180 mJ/pulse). A total of 4500 pulses were focused on the sample with a quartz lens through a calcium fluoride window. After the irradiation, additional rotational–vibrational bands of methane and acetylene were identified in the gaseous phase while the concentration of $H_2^{16}O$ increased (Fig. 2.2, curve c). Direct photocatalytic conversion of CO_2 into CH_4 a C_2H_2 occurred on the surface of

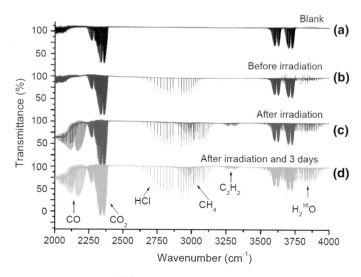

Fig. 2.2 Comparison of the **a** $C^{16,16}O_2$ spectrum with **b** the spectrum of the gas phase in the mixture of $Ti^{18}O_2$ (sample T200) and $C^{16,16}O_2$, **c** ditto after the irradiation with 4500 pulses of the XeCl laser, 308 nm, pulse width 28 ns, energy 180 mJ and spectrum **d** ditto after 75 h in dark at 30 °C Reprinted (adapted) with permission from Civiš et al. (2011). Copyright (2019) American Chemical Society

illuminated $Ti^{18}O_2$. Part of the unfocused laser radiation passed through the gaseous CO_2, which was partially broken down into CO. This molecule is identified by a rotational–vibrational band in the region of 2050–2200 cm^{-1} (Fig. 2.2, panel c–d). The sample was stored at temperature of 30 °C, and the spectra were measured after 50 and 75 h without UV-laser irradiation. A rotational–vibrational band of water (3600–3800 cm^{-1}) increased in intensity. The occurrence of $H_2^{16}O$ in the gaseous phase over the CO_2/TiO_2 interface, both in dark and upon illumination, deserves a special attention. All the main absorption rotational–vibrational lines belonged to $H_2^{16}O$, (the lines of $H_2^{18}O$ appeared only in their natural isotopic abundance). Concerning the water molecules which are released or created, no exchange of oxygen atoms ^{18}O from the solid phase ($Ti^{18}O_2$) took place.

The comparison of the reference spectra of methane and acetylene with the spectrum measured after the UV-laser irradiation of the $Ti^{18}O_2$ (T200) surface in a cell with $C^{16,16}O_2$ (b) is given in Fig. 2.3.

Carbon monoxide was created by laser irradiation in the cell (breakdown of CO_2); part of the spectrum is in Fig. 2.4. A small amount of $C^{16}O$ was generated from the gaseous CO_2. $C^{18}O$ was not generated in the cell, and the amount of $C^{16}O$ remained constant after 75 h.

There have already been studies on the adsorption of CO (Suriye et al. 2007). The rovibrational bands (2192 and 2209 cm^{-1}) were assigned to CO bonded to Ti^{4+} atoms in the lattice structure with a binding energy 17 kcal mol^{-1} (Thompson et al. 2003).

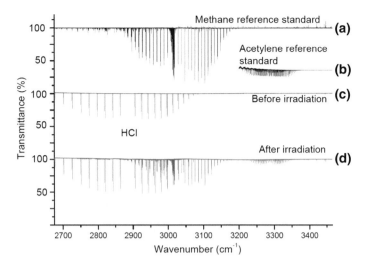

Fig. 2.3 Rreference spectra of **a** methane and **b** acetylene compared with **c** the sample without irradiation and **d** the sample after irradiation with 4500 pulses from the XeCl laser, 308 nm, pulse width 28 ns, energy 180 mJ. Reprinted (adapted) with permission from Civiš et al. (2011). Copyright (2019) American Chemical Society

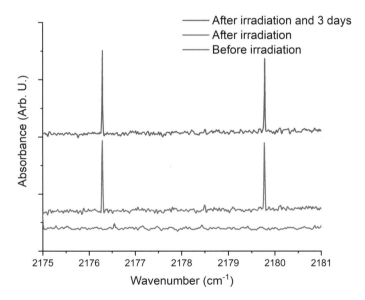

Fig. 2.4 $C^{16}O$ rotational–vibrational lines in the spectrum before the $Ti^{18}O_2$ (T200) irradiation (blue), after the irradiation with 4500 pulses from the XeCl laser (red) and after 75 h (black). Reprinted (adapted) with permission from Civiš et al. (2011). Copyright (2019) American Chemical Society

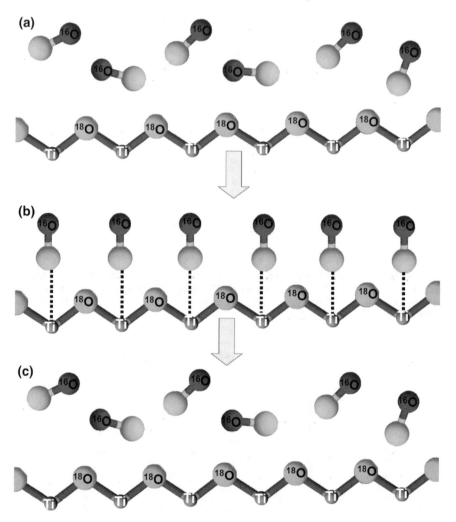

Fig. 2.5 Carbon monoxide bonding directly to titanium atoms on the surface of the crystalline $Ti^{18}O_2$. Since CO bonds straight to the titanium atom, no oxygen atom exchange is possible. After the laser irradiation or thermally at low pressure, the Ti–CO complex breaks back into $C^{16}O$ and $Ti^{18}O_2$. Reprinted (adapted) with permission from Civiš et al. (2011). Copyright (2019) American Chemical Society

The binding of CO to the Ti^{4+} (Fig. 2.5b) stands behind the fact that the oxygen isotope exchange does not occur between the carbon monoxide and the $Ti^{18}O_2$.

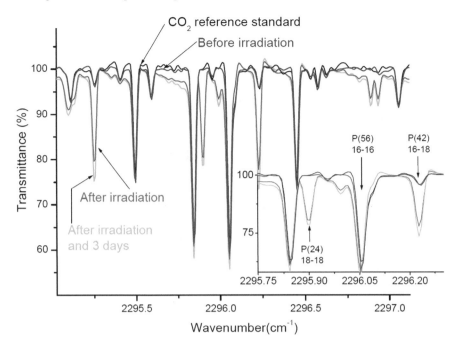

Fig. 2.6 Several rotational–vibrational lines of the ν_3 band of $C^{16,16}O_2$ in the spectral range 2295–2298 cm^{-1} (reference spectrum—black) together with the spectra of carbon dioxide (2 Torr) measured immediately after irradiation by the XeCl laser (red), and after 75 h (green). The absorption cell contained 0.8 g of $Ti^{18}O_2$ in the powder form (T200) spread over the walls of the glass reaction cell. Reprinted (adapted) with permission from Civiš et al. (2011). Copyright (2019) American Chemical Society

In the case of the reaction of $C^{16,16}O_2$ with $Ti^{18}O_2$ (T200), the spectral intensity of individual rotational–vibrational lines of CO_2 could be observed in the ν_3 region of the spectral band where the spectra of all three isotopologues ($C^{18,18}O_2$, $C^{18,16}O_2$, $C^{16,16}O_2$) overlap (Sorescu et al. 2011a). Figure 2.6 shows the reference spectrum of $C^{16,16}O_2$ (black) together with the rotational–vibrational transitions of individual isotopologues of carbon dioxide from the measured spectra. In the non-irradiated mixture of the T200 $Ti^{18}O_2$ and $C^{16,16}O_2$ measured directly after the cell was filled, no additional lines corresponding to the individual rotational–vibrational transitions of either $C^{18,18}O_2$ or $C^{18,16}O_2$ molecules were found (Fig. 2.6, blue curve) and the resulting spectrum fully corresponds to the spectrum of the pure $C^{16,16}O_2$.

Only after the irradiation of the sample with 4500 pulses by the UV laser, the $C^{18,16}O_2$ and $C^{18,18}O_2$ molecules appear in the gaseous phase (Fig. 2.6, red curve). Their concentration grew with time (Fig. 2.6, green curve). The individual lines in the spectra have been fitted and quantified. Thus calculated concentrations of individual isotopologues showed an incremental jump (after the laser irradiation) and a moderate increase (instead of decrease which one would intuitively expect) of the $C^{16,16}O_2$ isotopologue. The isotopologue $C^{18,18}O_2$ remained constant after the irradiation, and the isotopologue $C^{18,16}O_2$ showed a slight concentration increase over time in the mixture.

2.3 Spontaneous Oxygen Exchange Between $Ti^{18}O_2$ and $C^{16,16}O_2$

In the course of the experiments, it turned out that the oxygen atom exchange can take place even without the presence of light, and henceforth, we called the process a 'spontaneous oxygen exchange'. In order to verify this finding, we prepared $Ti^{18}O_2$, which was then, in the dark inside a high-resolution spectrometer, tested for the creation of the possible isotopologues of CO_2 ($C^{18,18}O_2$, $C^{18,16}O_2$ and $C^{16,16}O_2$) from the interaction with $C^{16,16}O_2$ with $Ti^{18}O_2$. The spectra were recorded by a high-resolution FTIR Bruker IFS 125 HR spectrometer, and the sample cell was directly connected to an outside inlet and outlet of gas for faster initiation of the reaction. The experimental set-up is shown in Fig. 2.7.

TiO_2 heated to a temperature of 200 °C (sample T200) showed low oxygen exchange activity. The process of isotope exchange and the adsorption of CO_2 are to a large extent influenced by the presence of water and OH groups on the surface of the TiO_2 (Sorescu et al. 2011b). After the irradiation of the sample, the $C^{18,16}O_2$ and $C^{18,18}O_2$ concentrations in the gaseous phase increased rapidly and then a slow process of spontaneous isotope exchange took place in the dark. A very interesting phenomenon is the release of adsorbed CO_2, which is almost completely composed

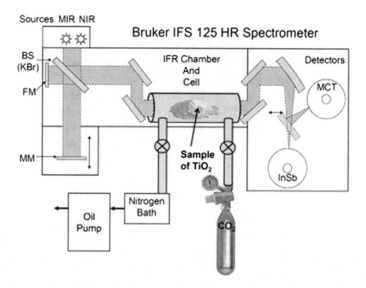

Fig. 2.7 High-resolution FT-IR laboratory set-up with the Bruker IFS 125 HR spectrometer

of C^{16,16}O$_2$ and the parallel process of the release of water from the surface of the sample, which is composed of H$_2$^{16}O. The release of water can be explained on the basis of the breakdown of the carbonic acid bonded to the surface of TiO$_2$:

$$H_2CO_3 \rightarrow CO_2 + H_2O \qquad (2.1)$$

As at the beginning of the experiment, H$_2$^{18}O is adsorbed on the surface of Ti^{18}O$_2$, and the carbonic acid is composed of a mixture of ^{16}O and ^{18}O. The increased concentration of H$_2$^{16}O in the gaseous phase above the TiO$_2$ surface after 75 h can be explained on the basis of the breakdown of the carbonate complex and the spontaneous isotope exchange reaction taking place in the dark:

$$H_2{}^{18}O + C^{16,16}O_2 \rightarrow C^{18,16}O_2 + H_2{}^{16}O \qquad (2.2)$$

This isotope exchange reaction is known and has the effect of enriching the carbon dioxide with the oxygen isotope ^{18}O. The effect of the spontaneous exchange of oxygen atoms between CO$_2$ and water has already been studied in several earlier papers. Mills and Urey (1940) in their 1940 paper report experimental results for the exchange in solutions with several pH values ranging from the acidic to the alkaline region. Of interest are mildly alkaline media, because they exhibit the prevalence of bicarbonate ions. The authors have shown in their study that the oxygen exchange in both acidic and mildly alkaline environments proceeds via the formation of a H$_2$CO$_3$ (as a direct hydration of CO$_2$ product). The overall mechanism of the isotope exchange process in such solutions has been explained in Suriye et al. (2007) and Lee et al. (2011). The oxygen isotope exchange between CO$_2$ and water adsorbed on Al$_2$O$_3$ or Fe$_2$O$_3$ was studied by Sorescu et al. (2011b). The rate constants of the spontaneous exchange in the solution are strongly temperature dependent. The mechanism of the possible processes taking place on the TiO$_2$ surface is shown in Fig. 2.8.

The isotopic exchange effects can be monitored either by observation of the envelope of the individual isotope absorption rotational–vibrational bands or by observation of the individual rotational–vibrational transitions of the C^{16,16}O$_2$, C^{18,16}O$_2$ and C^{18,18}O$_2$ molecules. Figure 2.9 (black curve) depicts a portion of the reference spectrum of the fundamental ν_3 band of carbon dioxide around 2300 cm^{-1}. The other traces depict the spectra of the gas phase in a cell loaded with a Ti^{18}O$_2$ solid (0.8 g, sample T450). After the annealing, 2 Torr of C^{16,16}O$_2$ were transferred to the sample. Blue line shows a gas phase spectrum recorded immediately after the addition of C^{16,16}O$_2$. Green and red lines show the spectra observed after 15 h (green) and 50 h (red), respectively.

The blue spectrum taken immediately (a few seconds) after the addition of C^{16,16}O$_2$ shows that the oxygen exchange between our TiO$_2$ T450 sample and CO$_2$ is a very fast process. The depicted spectrum shows the rovibrational lines of C^{18,16}O$_2$ together with the lines of C^{18,18}O$_2$.

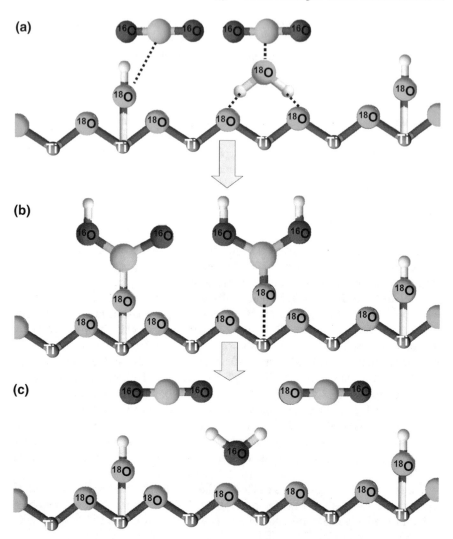

Fig. 2.8 Diagram of the isotope exchange on the surface of crystalline $Ti^{18}O_2$. The surface layer contains Ti^{4+} atoms onto which the ^{18}OH groups and $H_2{}^{18}O$ are bonded. The gaseous carbon dioxide reacts with water and OH groups, creating a hydrogen and dihydrogen complex bonded to the Ti atoms. This complex breaks apart after the laser irradiation or thermally at low pressure, back into molecules of $H_2{}^{16}O$, $C^{16,16}O_2$ and partially also into $C^{18,16}O_2$. Reprinted (adapted) with permission from Civiš et al. (2011). Copyright (2019) American Chemical Society

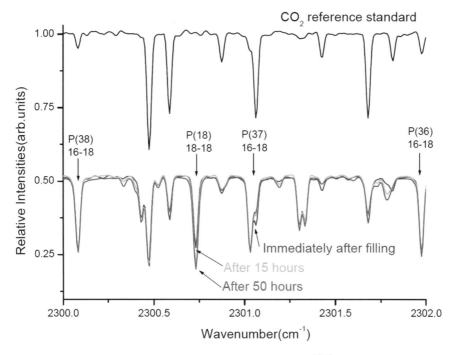

Fig. 2.9 Several rotational–vibrational lines of the ν_3 band of C^{16,16}O$_2$ in the spectral range 2300–2302 cm^{-1}. The black line shows a reference C^{16,16}O$_2$ spectrum. The blue, green and red line show a spectrum of 2 Torr of C^{16,16}O$_2$ in free contact with a Ti^{18}O$_2$ mineral immediately after their contact and after 15 and 50 h, respectively. The cell contained in total 0.8 g of powdered Ti^{18}O$_2$. Reprinted (adapted) with permission from Civiš et al. (2011). Copyright (2019) American Chemical Society

Figures 2.9 and 2.10 show the exchange monitored on the overtone bands in the 3500–3800 cm^{-1} region. The results are the same as in Fig. 2.7.

2.4 Section Summary

The isotope exchange has been used, as described above, in the reverse order relative to the majority of the published work, i.e. using ^{16}O-labelled gas and ^{18}O-labelled mineral. Our team, Civiš et al. (Liao et al. 2002) have demonstrated how isotopically relatively pure Ti^{18}O$_2$ samples can be prepared from various Ti-precursors via hydrolysis using heavy-oxygen water, H$_2$^{18}O. Both anatase and rutile forms have been synthesized using either TiF$_4$ or TiCl$_4$ precursors, and formation of pure Ti^{18}O$_2$ isotopologues has been demonstrated based on an analysis of Raman spectra (Bogdanoff and Alonso-Vante 1994). Furthermore, the reactions of C^{16,16}O$_2$ with Ti^{18}O$_2$ anatase under both thermal-excitation and photoexcitation conditions have been con-

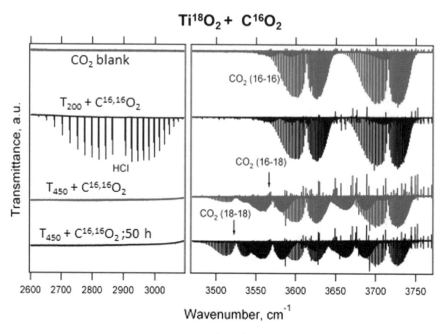

Fig. 2.10 High-resolution FTIR spectra of the gas phase in the exchange experiment between $Ti^{18}O_2$ and $C^{16,16}O_2$. The upper red spectrum shows a $C^{16,16}O_2$ standard spectrum. The upper black spectrum then shows the T200 $Ti^{18}O_2$ sample and $C^{16,16}O_2$ after 30 min of mutual contact. Still, only $C^{16,16}O_2$ is visible. The lower red spectrum shows the gas phase composition of the $C^{16,16}O_2$ and T450 $Ti^{18}O_2$ mixture. $C^{18,16}O_2$ and $C^{18,18}O_2$ are clearly visible. The lower black spectrum then shows the same sample after 50 h in the dark at ambient temperature. The assignment of characteristic bands is marked by arrows. Reproduced from Kavan et al. (2011) with permission from the PCCP Owner Societies

sidered (4, 37), and the formation of methane and acetylene was observed in the presence of water upon UV irradiation. Following the chemical evolution of CO_2 on the oxide surface, it has been demonstrated that $Ti^{18}O_2$ anatase samples annealed in vacuum at 450 °C can exchange oxygen with gaseous $C^{16,16}O_2$ leading to formation of $C^{18,18}O_2$ as the major product with a minor content of $C^{18,16}O_2$ (Liao et al. 2002). Isotopic oxygen exchange has been suggested to take place by adsorption of gaseous $C^{16,16}O_2$ molecules at surface oxygen defects where they bond to a surface ^{18}O and form bidentate CO_3^{2-} species with subsequent thermal release of either $C^{18,16}O_2$ and subsequently $C^{18,18}O_2$ species (Liao et al. 2002). A similar CO_3^{2-} intermediate was assumed by Yanagisawa and Sumimoto (Yanagisawa and Sumimoto 1994) in their study of the oxygen exchange of $C^{18,18}O_2$ with vacuum-annealed $Ti^{16}O_2$ powders. The authors, unfortunately however, did not present detailed overall mechanism of the exchange.

2.5   On the Mechanism of the Spontaneous Oxygen Exchange Between $C^{18,18}O_2$ and $Ti^{16}O_2$ (101) Anatase

Following the work of Yanagisawa and Sumimoto (Yanagisawa and Sumimoto 1994) then, we explored the oxygen atom exchange between $C^{18,18}O_2$ and $Ti^{16}O_2$. The isotope labelling is only a tool for the study of the interaction, and the yielded results were qualitatively the same. This 'classical' model of the labelled gas and non-labelled mineral was, however, used for theoretical computations of the interaction energies and for the explanation of the mechanism of the exchange.

An important feature of the exchange process is the formation of oxygen vacancies (V_O) in the mineral surface layer (Jimenez et al. 1993; Henderson 1995; Yanagisawa 1995; Kalamaras et al. 2009). The most stable form of anatase in the context of the formation of these vacancies is anatase (101) surface, whose tendency to form vacations is the smallest (Kalamaras et al. 2009). This has been observed, among others, by comparison with the (110) rutile surface. This disparity has been attributed to a relatively lower stability of the fourfold Ti^{3+} sites on the anatase (101) surface in contrast to the fivefold Ti^{3+} centres of the rutile (110) surface.

Further, the density functional theory (DFT) calculations by Henderson et al. (Henderson 1995) have shown that the energy difference between the formation of a bulk vacancy and a surface defect in anatase is about 0.5 eV. Even more recently, scanning tunnelling microscope (STM) analysis of the oxygen surface defect distribution at various temperatures has been reported (Jimenez et al. 1993). That study shows at temperature above 200 °C, surface defects tend to migrate in the crystal towards subsurface layers. The authors also report reappearance of some oxygen defects indicating a to and fro migration of the defects through the crystal. It has been pointed out by some (Scheiber et al. 2012) that the performed DFT calculations did not take this migration process and multiple-layer structure into account and that it is thus difficult to reconcile the calculations with a thermal equilibrium between the surface and subsurface defects.

The creation of oxygen vacancies and Ti^{3+} centres is not limited to microcrystalline or powdered TiO_2. It was shown that these point defects may be created on a nanocrystalline TiO_2 as well using relatively mild conditions. The Praserthdam research group have shown that such surface defects can be created by changing the amount of oxygen during calcination (Brenninkmeijer et al. 2003) or changing the water:alkoxide ratio in the case of the sol-gel synthesis (Scheiber et al. 2012), both at mild temperatures.

In samples prepared using the latter method, the obtained temperature-programmed desorption (TPD) spectra of CO_2 indicate the presence of broad peaks at 175 and 200 K. These peaks were assigned, respectively, to the adsorption of CO_2 at regular Ti^{4+} and at Ti^{3+} vacancy sites (Hadjiivanov et al. 1997). This assignment was done in direct analogy with the 170 and 200 K TPD peaks recorded for CO_2 adsorbed on oxidized and reduced (110) rutile (Mills and Urey 1940). Given the close correspondence between the TPD data on the anatase and rutile surfaces, it can be expected that the binding energies of CO_2 on the anatase (101) surface closely resemble those on the rutile (110) surface. We note that zero-coverage adsorption energies of 11.59 and 12.90 kcal/mol have been determined for CO_2 on the fully oxidized and reduced rutile (110) surfaces, respectively (Mills and Urey 1940).

We have now extended our previous theoretical investigation (Mills and Urey 1940) of the adsorption of CO_2 on the oxidized and defective anatase (101) surface by including oxygen exchange between gaseous CO_2 and surface oxygen atoms. As shown in previous studies of adsorption of CO_2 on the rutile and anatase (Mills and Urey 1940) surfaces, an accurate description with almost quantitative agreement between experiments and theoretical predictions can be obtained if corrections for long-range dispersion interactions are included in the DFT calculations.

It was concluded earlier (Sato 1987; Liao et al. 2002) that oxygen atom exchange between gaseous CO_2 and the anatase surface is mediated by the formation of a carbonate species upon CO_2 adsorption at a surface V_O defect site. In order to test this hypothesis, we first performed analyses of various adsorption configurations of CO_2 near or directly at V_O sites on the anatase (101) surface layer. From the obtained possible configurations, we aimed at selecting possible candidate conformations/configurations for the formation of CO_3-like species. The considered CO_2/TiO_2 configurations are shown on the left side of Fig. 2.11.

As is shown in Fig. 2.11, CO_2 adsorbs at the V_O site in bent or linear configurations (L(1), L(2), B(1) or B(2)). In the bent configurations, covalent bonds are formed during the adsorption (chemisorption). On the other hand, in the linear configurations, the molecule is only physisorbed and no true covalent bonds are formed.

The adsorption energies E_{ads} were calculated as

$$E_{ads} = \frac{\left(n E_{CO_2} + E_{slab} - E_{CO_2+slab}\right)}{n} \tag{2.3}$$

where E_{CO_2} is the energy of the CO_2 molecule at its optimized gas phase geometry, n represents the number of adsorbate molecules in the simulation cell, E_{slab} is the total energy of the slab, and E_{CO_2+slab} is the total energy of the adsorbate/slab system. In this sign convention, positive adsorption energies correspond to stable configurations. Isolated CO_2 molecule energy was calculated in a cubic cell with 12Å sides.

In addition to the four configurations discussed above and which we reported previously by Rothman et al. (2009), we present a new binding configuration C(**0**) with the CO_2 bonded to the surface by 14.6 kcal/mol. This corresponds to a bent CO_2 molecule adsorbed at V_O site in a laying down configuration with the formation of two new bonds, one with a surface Ti(5f) atom and a second with a nearby bridging O(2f) atom. The C(0) intermediate configuration symmetrically resembles a CO_3-like species bonded to two neighbouring Ti(5f) atoms in the surface layer. As shown in the representation C(0), the CO_3-like entity has a D_{3h} symmetry and all the three oxygen atoms are chemically equal (the two original CO_2 oxygen and the one relevant O(2f) lattice oxygen). This intermediate carbonate species is ideal for promoting the isotope exchange.

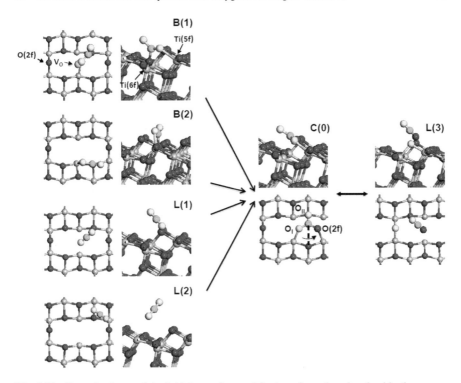

Fig. 2.11 General scheme of the initial, transient and final configurations involved in the oxygen exchange mechanism between CO_2 and a defective anatase (101) surface. Left part of the figure depicts four adsorption configurations (side and top view) of CO_2 at a V_O site. L in L(1) and L(2) stands for linear and B in B(1) and B(2) stands for bent which indicates the possible bending of the CO_2 molecule. In the B case, the model also shows the formation of new bonds between CO_2 and Ti(5f) and O(2f) surface atoms. The central panels show the C(0) configuration. The right side panel depicts views of a carbon dioxide molecule with a new oxygen atom adsorbed in L(3) linear configuration. The L(3) configuration is symmetrically equivalent to L(1). Surface Ti an O atoms are depicted in grey and red, respectively, ^{16}O from CO_2 is yellow and the carbon atom is green. In the top left model structure, the light blue circle represents the V_O site. O_I in the C(0) configuration denotes the O atom from the CO_2 molecule closest to the surface, which is subsequently bonded to the Ti(5f) atom while the O_{II} indicates the other, non-interacting O atom of the molecule. Reprinted (adapted) with permission from Civiš et al. (Sorescu et al. 2014). Copyright (2019) American Chemical Society

As shown in Fig. 2.12, the energy profiles for L(**1**) → C(**0**) and L(**3**) → C(**0**) are essentially equivalent. The barriers for the entry L(**1**) → C(**0**) and for the exit C(**0**) → L(**3**) steps are calculated to be 12.5 and 11.4 kcal/mol, respectively. These barriers can be easily overcome by thermal excitation under the experimental conditions (Bogdanoff and Alonso-Vante 1994).

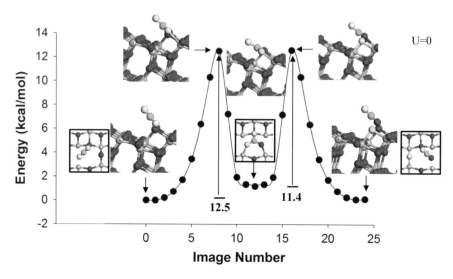

Fig. 2.12 Minimum total energy change landscape for the oxygen atom exchange between CO_2 and anatase. The reaction proceeds through L(1), C(0) and L(3) states in order as determined from PBE-TS DFT calculations with the Hubbard U correction term 0 eV. Reprinted (adapted) with permission from Civiš et al. (Sorescu et al. 2014). Copyright (2019) American Chemical Society

The results provided by the PBE-TS DFT calculations give binding energies consistent with the experiments (Brenninkmeijer et al. 2003). What is more, the calculated barriers in the L(1)-C(0)-L(3) sequence have such values that the oxygen exchange can be initiated by thermal excitation at room temperature, which again is in agreement with experiments (Bogdanoff and Alonso-Vante 1994). As seen in Fig. 10 in Sorescu et al. (Sorescu et al. 2014) for $U > 2.5$ eV, the computed barriers are large (20.4–24.2 kcal/mol) for CO_2 to leave the carbonate-like intermediate. Such a barrier makes the spontaneous exchange very improbable. Only in cases where $U < 2.5$ eV, the barriers are low enough to allow for thermally mediated isotope exchange at room temperature.

For $U = 0$ eV, the calculated activation energies for the oxygen exchange mechanism between CO_2 and solid TiO_2 are small enough ($E_a < 12.5$ kcal/mol) to support initiation of the oxygen exchange at ambient temperature.

A final comment pertains to the oxygen exchange mechanism. As shown in this study, the intermediate C(0) carbonate-like species can be formed at the V_O defect site. In such case, the exchange occurs with the involvement of the bridging O atom from the lattice structure. Were the defect site rooted to the spot, the reaction rate would be severely limited (for sooner or later the surface area of the crystal would have reached equilibrium with the gas reactant). This is in contrast to experimentally observed results (Bogdanoff and Alonso-Vante 1994). The resolve of this discrepancy lies in the mobility of the vacancies. Scheiber et al. (Jimenez et al. 1993) have shown that following equilibration, a continuous exchange of atoms and defects between the surface and subsurface layers creates new sites for the exchange. The exchange capacity of the mineral is in this way greatly increased.

2.6   Oxygen Atom Exchange Between $C^{18,18}O_2$ and $Ti^{16}O_2$ Nanoclusters

Seeing that the oxygen atom exchange works well with common nanocrystalline TiO_2 anatase, we investigated the rate of the formation of $C^{16,16}O_2$ from the $C^{18,18}O_2$–$Ti^{16}O_2$ system using a high-surface titania synthesized from isopropoxide. Surprisingly, the formation of the $C^{16,16}O_2$ from $C^{18,18}O_2$ was significantly faster than in the case of common anatase. The rate of the formation is strongly dependent on the calcination temperature, growing with the increasing temperature.

Titania has complex surface properties, whose understanding, however, is essential for the understanding of its surface chemistry. There exist various oxygen defect sites in the mineral which influence the rate of the exchange. These can naturally exist both in the bulk of the crystal and on its surface. Mild reduction conditions may create artificial vacancies as well as Ti^{3+} sites and both these defects can trap electrons. These sites also play a role in the n-doping of the mineral (Wendt et al.

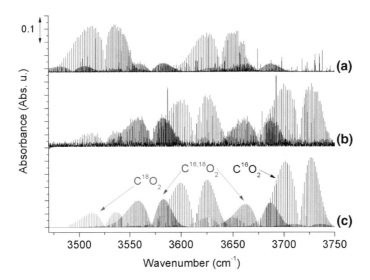

Fig. 2.13 Infrared spectra of $C^{18,18}O_2$, $C^{18,16}O_2$ and $C^{16,16}O_2$ $\nu_1 + \nu_3$ and $2\nu_2 + \nu_3$. Spectrum **a** shows $C^{18,18}O_2$ in the absence of a mineral surface. Spectrum **b** shows $C^{18,18}O_2$ after 1000 s of being in contact with anatase annealed at 400 °C. Spectrum **c** finally shows the $C^{18,18}O_2$ (red), $C^{18,16}O_2$ (blue) and $C^{16,16}O_2$ (black) vibrational bands used for the assignment. The bottom spectrum was simulated using the Winproof program and the HITRAN database. Reprinted from Civiš et al. (2013), Copyright (2019), with permission from Elsevier

2008; Di Valentin et al. 2009). Both these defects can also migrate and when at the surface, the missing oxygen creates a point defect and three undercoordinated (fivefold) Ti ions. The two remaining electrons fill the empty orbitals of the Ti^{4+} ions (formally). The undercoordinated surface sites influence massively the reactivity of the mineral as a whole. In an ideal (101) anatase, both fivefold (Ti(5c)) and sixfold (Ti(6c)) sites are present. When reduced in mild conditions, fourfold (Ti(4c)) coordinated sites are created as well. It is these undercoordinated Ti(4c) sites that are mostly responsible for the faster interaction (Qu and Kroes 2007; Sorescu et al. 2011a). The surfaces of the common crystalline anatase and the quasi-amorphous TiO_2 made from isopropoxide are different from the surface defect density point of view, and consequently, the activity of the quasi-amorphous TiO_2 was found to be significantly greater than that of the common anatase.

The quasi-amorphous high-surface TiO_2 was prepared by the hydrolysis of isopropoxide. The precipitated TiO_2 was then dried at 60 °C in air. Both samples (TiO_2 from isopropoxide, designated NANO and common TiO_2 from $TiCl_4$ designated A) were annealed at temperatures up to 450 °C, and at each temperature, the surface area was measured as shown in Figs. 2.14 and 2.15. The surface was determined by measuring nitrogen adsorption isotherm at 77 K (ASAP 2010, Micrometrics) and their subsequent treatment in the range of relative pressures 0.075–0.25 using the Brunauer–Emmett–Teller (BET) method.

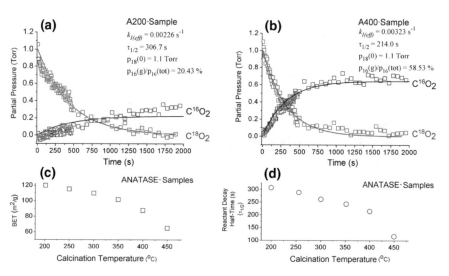

Fig. 2.14 Anatase surface area and exchange activities. Panels A and B show the time evolution of the system $C^{18,18}O_2$–$Ti^{16}O_2$ for the sample calcined at 200 and 400 °C. As time progresses, the concentration of $C^{18,18}O_2$ decreases and the concentration of $C^{16,16}O_2$ increases. Panel C shows the surface area as it decreases with the increasing calcination temperature and panel D shows the half-lives of the $C^{18,18}O_2$ concentration in the system versus the temperature of calcination of the material. Reprinted (adapted) with permission from Civiš et al. (2015). Copyright (2019) American Chemical Society

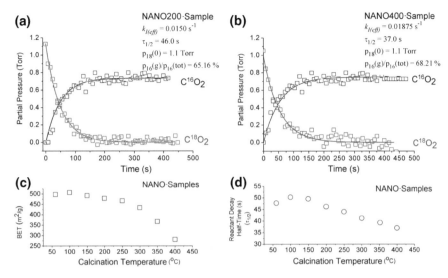

Fig. 2.15 NANO sample surface area and exchange activities. Panels A and B show the time evolution of the system $C^{18,18}O_2$–$Ti^{16}O_2$ for the sample calcined at 200 and 400 °C. As time progresses, the concentration of $C^{18,18}O_2$ decreases and the concentration of $C^{16,16}O_2$ increases. Panel C shows the surface area as it decreases with the increasing calcination temperature and panel D shows the half-lives of the $C^{18,18}O_2$ concentration in the system versus the temperature of calcination of the material. Reprinted (adapted) with permission from Civiš et al. (2015). Copyright (2019) American Chemical Society

The catalysts, calcined at different temperatures, were put into contact with $C^{18,18}O_2$, and its concentration as well as the concentration of $C^{18,16}O_2$ and $c^{16,16}O_2$ was monitored again. The corresponding example spectra are shown in Fig. 2.13.

The $C^{18,16}O_2$ attained a low concentration and therefore was not plotted in the final figures. Figures 2.14 and 2.15 show in panels A and B the concentration of $C^{18,18}O_2$ and $C^{16,16}O_2$. The data were fitted with a first-order rate equation, as described in the section below, and half-lives of the CO_2 species were determined. From these, the comparison of effectivity could be made between the A samples (common anatase) and the NANO samples (quasi-amorphous high-surface anatase). It is clear from the conversion curves and the rate constants that the NANO samples are much more efficient in the exchange than the common anatase samples. As noted in the introduction, the exchange reaction occurs at either the bridging O atoms or the Ti(4c) uncoordinated sites. The process of the exchange can be described by a two-step reaction. Given the nature of the process and the low concentration of the intermediate $C^{18,16}O_2$ molecule, the kinetic fit may be simplified and the reaction may be described as a one-step process leading from $C^{18,18}O_2$ directly to $C^{16,16}O_2$. While this approach is not correct from the point of view of the order of the reaction, it is an easy convenient tool for the qualitative description of the behaviour of the different samples. The results are shown in Figs. 2.14 and 2.15 in panels D.

For both samples, the reaction half-lives and the BET surface area decreases. The decrease of the surface area at sufficiently high temperature may be explained by the sintering of the material (Kavan et al. 1996). On the other hand, the increase in activity towards the interaction with CO_2 may be ascribed to the increased density of oxygen vacancies, which are created in the material through annealing. The annealing perhaps creates the undercoordinated Ti(4c) sites as well through the temperature-induced dismissal of oxygen atoms from the surface. These effects contradict each other as higher surface area provides more opportunity for the presence of the active sites. It seems, however, that the production of the active defects is the prevailing process in this case. It should be noted, that the chemical reactivity of both samples (the crystalline and NANO-) is similar and remains unchanged.

It was shown in the section above that in the course of the oxygen atom exchange, the CO_2 molecule is first physisorbed on the surface of the mineral and to do so, needs to pass a barrier of about 10.9 kcal mol^{-1} (for anatase (101), the Hubbard U term 2.5 eV). After that, the CO_3-like structure is formed and all its oxygen atoms are symmetrically and chemically degenerate. The CO_2 can be released again, for which it needs to scale an energy barrier of about 16.5 kcal mol^{-1} (the Hubbard U term 2.5 eV).

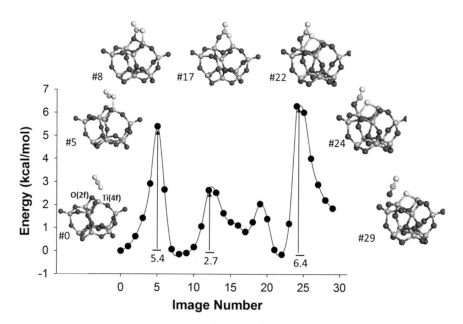

Fig. 2.16 Minimum energy pathway for the oxygen atom exchange on quasi-amorphous titanium dioxide. The barriers for the adsorption and desorption are considerably lower, namely 5.4 and 6.4 kcal mol^{-1} (with the Hubbard U term 2.5 eV). A full description of the mechanism can be found in Civiš et al. (2015). Reprinted (adapted) with permission from Civiš et al. (2015). Copyright (2019) American Chemical Society

For the purpose of the investigation of the energy profile of the exchange on the quasi-amorphous NANO sample, we used a $(TiO_2)_{10}$ cluster described by Marom et al. (2012). This cluster was chosen because it contains a significant amount of Ti(4c) undercoordinated sites, which affect the oxygen exchange rate. Minimum energy pathway for the oxygen exchange was computed on this cluster and is shown in Fig. 2.16.

We, therefore, see that the barriers in the case of the quasi-amorphous TiO_2 are smaller and the reaction can therefore proceed faster in ambient conditions where only thermal motion is available as a source of energy. This is the reason behind the faster oxygen atom exchange on the high-surface NANO sample. Having now the explanation of the mechanism complete, we could move forward to the exploration of the activity of other minerals.

2.7   Breathing Minerals

After the exchange was confirmed and studied between titanium dioxide in its various forms and CO_2, the question arose whether this property is exclusive to the TiO_2 or whether other minerals interact with their surroundings as well. We selected several natural and synthetic minerals and tested their effectivity in the exchange experiments in the same manner as in the previous studies. In each experiment, a mineral phase was put into contact with gaseous $C^{18,18}O_2$ and the high-resolution gas phase infrared spectra of $C^{18,18}O_2$, $C^{18,16}O_2$ and $C^{16,16}O_2$ were monitored.

2.7.1   Synthetic and Natural Anatase

In all these experiments, the isotopic exchange was monitored using Fourier transform infrared spectroscopy and fundamental, overtone or combination bands were monitored for each molecule. Figure 2.17 shows a representative spectrum of the fundamental v_3 and the $v_1 + v_3$ and $v_1 + 2\,v_3$ combination bands (Sandford et al. 1991). The upper spectrum (red) shows $C^{18,18}O_2$ without the presence of any solid phase mineral. The lower spectrum (black) shows the mixture composition after the $C^{18,18}O_2$ being left in contact with $Ti^{16}O_2$ for 15 000 s. The TiO_2 was calcined at 450 °C prior to the experiment. All the initial $C^{18,18}O_2$ was converted to $C^{16,16}O_2$, and no $C^{18,18}O_2$ is visible in the sample anymore.

Initially, synthetic and natural anatases were tested. The natural anatase was obtained from Norway, from the Hordaland County. The synthetic anatase was synthesized at the J. Heyrovský Institute of Physical Chemistry by prof. Kavan and his team. Both samples were shown to exhibit exchange activity. The rates are different for each sample and are dependent on the calcination temperature. Figure 2.18 shows conversion curves for both minerals calcined at 100 and 450 °C. Panel A of Fig. 2.18 depicts conversion on synthetic anatase. It is clear that the half-life of $C^{18,18}O_2$ is

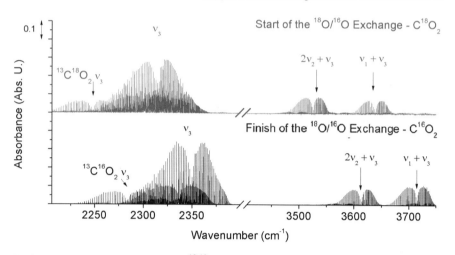

Fig. 2.17 Upper red spectrum shows $C^{18,18}O_2$ without the presence of any solid phase mineral. The lower spectrum (black) shows the mixture composition after the $C^{18,18}O_2$ being left in contact with $Ti^{16}O_2$ for 15 000 s. In the lower spectrum, all the initial $C^{18,18}O_2$ was converted to $C^{16,16}O_2$ and no $C^{18,18}O_2$ is visible in the sample anymore. Reprinted (adapted) with permission from Civiš et al. (2016b). Copyright (2019) American Chemical Society

shorter when the anatase is calcined at higher temperature. Panel B of Fig. 2.18 shows the data for the exchange on natural anatase. Here, the trend is even stronger as the calcination decreases the half-life over $12\times$. The half-lives for the natural anatase are overall longer.

The conversion curves (both the $C^{18,18}O_2$ decrease and the $C^{16,16}O_2$) were fitted with first-order rate equation to obtain pseudo-first-order rate constants. We shall call the constants pseudo-first-order since the process is by nature not a first-order kinetic process. However, the equation fits the conversion curves reasonably well and the obtained pseudo-first-order rate constants provide an easy tool for the comparison of the effectivity of the samples.

The mixed isotope oxide, $C^{18,16}O_2$, is a natural intermediate in the exchange. Its concentration, however, does not exceed 3% during the experiment. The exchange can therefore be simplified for the purpose of the fit as:

$$C^{18,18}O_2 \rightarrow C^{16,16}O_2 \tag{2.4}$$

The first-order rate equation for this process can be written as:

$$-\frac{dp_{18}(t)}{dt} = k_{I,\text{eff}} \times dp_{18}(t) \tag{2.5}$$

where $p_{18}(t)$ is the partial pressure of $C^{18,18}O_2$. Therefore, $p_{18}(t)$ can be expressed as:

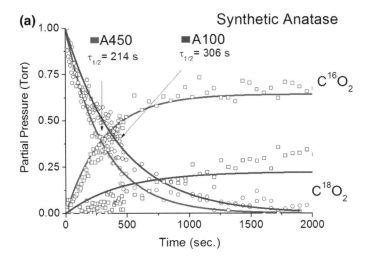

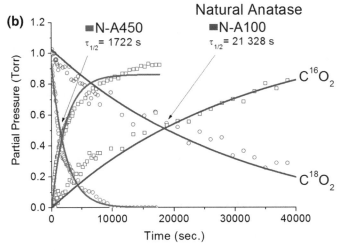

Fig. 2.18 Panel A shows the exchange on synthetic anatase at 100 and 450 °C. Panel B shows the exchange data for natural anatase. Reprinted (adapted) with permission from Civiš et al. (2016b). Copyright (2019) American Chemical Society

$$p_{18}(t) = p_0 \times e^{-k_{I,\text{eff}}t} \qquad (2.6)$$

where p_0 is the initial pressure of $C^{18,18}O_2$. In a similar way, the partial pressure of the product $C^{16,16}O_2$, $p_{16}(t)$, can be expressed as:

$$p_{16}(t) = p_0 \times \left(1 - e^{k_{I,\text{eff}}t}\right) \times \frac{p_{16}}{p_0} \qquad (2.7)$$

where p_{16} is the total partial pressure of $C^{16,16}O_2$ created in the reaction (the final measured value) and p_0 is the initial pressure of $C^{18,18}O_2$, i.e. the total pressure in the sample. This ratio is included because some of the produced or inserted CO_2 is adsorbed on the surface. The respective rate constants are shown in each figure at the conversion curves.

As is seen from the rate constants and from the conversion curves, the exchange rate is significantly boosted by calcination of the samples at high temperatures (450 °C compared to 100 °C). This is caused by the fact that vacancies form on the crystal surface at high temperatures and new adsorption defect sites are created.

2.7.2 Synthetic and Natural Rutile

Similar experiments with the oxygen exchange were performed on rutile. Two samples were again used: synthetic rutile (Bayer 5556), which has a large surface area and which was dried in vacuum and annealed at 100 °C, and natural rutile from Golčův Jeníkov, Czech Republic, which was calcined at 100 and 450 °C. The rutile was obtained from veins of orthoamphibolite, quartz veins of paragneisses and in the contact zone between amphibolite xenoliths and the granite from Přibyslavice.

As it turned out, all the rutile samples exhibit very fast oxygen exchange. The experimental data are shown in Fig. 2.19. The data were fitted with the pseudo-first-order rate equation, similar to the anatase exchange experiment. Panel A shows the exchange of oxygen between $C^{18,18}O_2$ and synthetic rutile Bayer 5556. The half-life of the isotopically labelled carbon dioxide is about 0.09 s! Panel B shows data for the natural rutile from Golčův Jeníkov. It is clear that annealing at 450 °C increases the rate of oxygen exchange by a factor of approximately 3.5.

Figure 2.20 shows the adsorption capacities of CO_2 for the tested minerals. It is shown here that the synthetic rutile has larger adsorption capacity for CO_2 compared to anatase both synthetic and natural. This can be again explained by the fine structure of the material. The oxygen exchange, as described in the previous chapters, is driven by the presence of oxygen vacancies in the crystal lattice. The large surface area of the material and the fine structure of the material mean that the sample is overall more heterogenous and contains more edges and crystal defects created at the surfaces of particles.

It is shown here on the case of rutile and anatase, that there are two fundamental effects acting in the same way (Sorescu et al. 2011b, 2012). The surface area is one of them and as it increases, the activity for the oxygen exchange increases as well. The other is the presence of oxygen defects such as vacancies. When the calcination temperature is increased, the surface density of the defects increases as well. The surface area, in the case of anatase, which is stable up to 700 °C, remains largely unchanged. As long as these effects are synergic, the annealing will always increase the oxygen exchange effectivity. But is it always so?

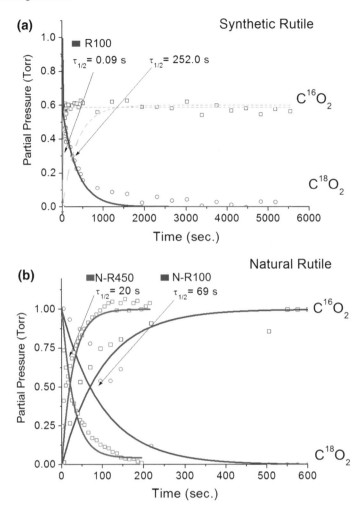

Fig. 2.19 Panel A shows the exchange of oxygen between $C^{18,18}O_2$ and synthetic rutile. Panel B shows similar data for natural rutile annealed at 450 °C (red) and 100 °C (blue). Reprinted (adapted) with permission from Civiš et al. (2016b). Copyright (2019) American Chemical Society

2.7.3 Montmorillonite and Clay from the Sokolov Coal Basin

The oxygen exchange on anatase and rutile shows that the process is not limited to one mineral phase. Would a different mineral exhibit the same properties? Montmorillonite or clays often contain TiO_2 as an admixture, but not as a crystal. Was the property dependent on the crystalline form on the mineral? To answer these questions, the portfolio of minerals was broadened. Since clays are very common minerals not only on Earth, but also Mars, and since they have many catalytic properties, whose role has been discussed, for example, in the context of the origin of life, the next

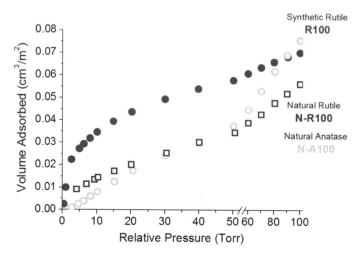

Fig. 2.20 Comparison of CO₂ adsorption capacities of the tested minerals—natural anatase, natural rutile and synthetic rutile. Reprinted (adapted) with permission from Civiš et al. (2016b). Copyright (2019) American Chemical Society

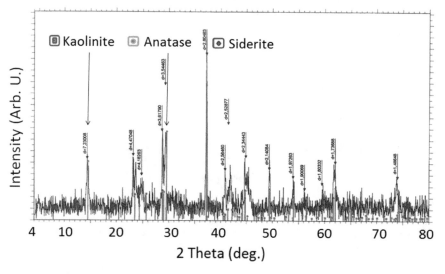

Fig. 2.21 X-ray diffraction analysis of clay from the Sokolov coal basin. The main components of the clay are kaolinite, siderite and anatase. Reprinted (adapted) with permission from Civiš et al. (2016b). Copyright (2019) American Chemical Society

tested minerals were montmorillonite and natural clays from the Sokolov coal basin in the Czech Republic.

To explore the oxygen isotopic exchange on clays, we collected samples in the Družba Quarry, Sokolov, Czech Republic. The clays contain TiO₂ and the exchange

Fig. 2.22 EDX analysis of elements in the clay sample. Al, Si and Ti were identified as the main elemental components (besides O). Reprinted (adapted) with permission from Civiš et al. (2016b). Copyright (2019) American Chemical Society

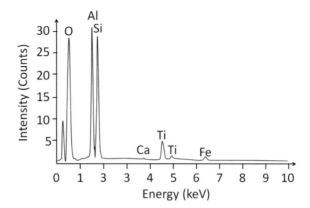

Table 2.1 Results of EDX analysis of the clay sample shown in atom %

Element	Content (at.%)	Error (at.%)
O	75.9	0.601
Mg	0.165	0.035
Al	11.146	0.091
Si	11.269	0.095
Ca	0.047	0.01
Ti	1.185	0.029
Fe	0.289	0.032

As seen in the plot above, Al, Si and Ti are the main elements in the sample (besides oxygen). Reprinted (adapted) with permission from Civiš et al. (2016b). Copyright (2019) American Chemical Society

was initially expected to work on the titanium dioxide only. Macroscopically, the samples are a silty clay with volcanogenic admixtures, bioturbed (from fossil soil) and permeated with coalificated plant roots. The sample was analysed by X-Ray diffraction analysis, and the resulting spectrum is shown in Fig. 2.21. Main components were revealed to be anatase, siderite and kaolinite. Elemental composition of the sample is shown in Fig. 2.22 and Table 2.1 as determined by EDX.

The results of the oxygen exchange between $C^{18,18}O_2$ and the Sokolov clay are shown in Fig. 2.23 panel A. In contrast to pure TiO_2, annealing at 450 °C leads to decreased oxygen mobility between the two compounds. The oxygen exchange is less effective and the half-lives are longer. This trend can be explained by the structure of the material. That the oxygen mobility is not dependent solely on the chemical environment and the oxygen vacancies in TiO_2 are shown in Fig. 2.23 panel B which illustrates that pure montmorillonite (synthetic, type K10, Sigma Aldrich) exhibits oxygen exchange as well. Since the synthetic montmorillonite does not contain any TiO_2, the exchange must take place through a different mechanism. The activity of the fine nanoparticles of the montmorillonite along with its complex

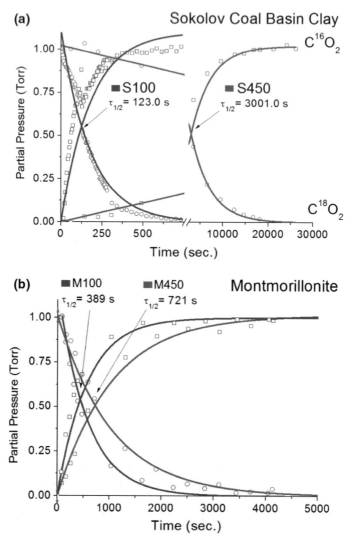

Fig. 2.23 Conversion curves for the oxygen exchange between $C^{18,18}O_2$ and montmorillonite (panel B) or the Sokolov clay (panel A), both annealed at 100 and 450 °C. Reprinted (adapted) with permission from Civiš et al. (2016b). Copyright (2019) American Chemical Society

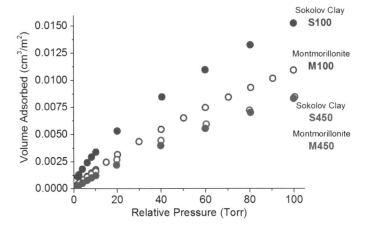

Fig. 2.24 Adsorption capacities of montmorillonite and Sokolov clay annealed at 100 and 450 °C. Samples annealed at 100 °C are shown in blue and samples annealed at 450 °C in red. Montmorillonite is shown by hollow circles while Sokolov clay is shown by full circles. Reprinted (adapted) with permission from Civiš et al. (2016b). Copyright (2019) American Chemical Society

surface structure provides environment for the oxygen atom exchange. Annealing at higher temperature (in our case 450 °C) results in sintration. During sintration of the material, the particles stick together and therefore the average particle size increases, which means that the surface area and the adsorption capacity both decrease. This hypothesis was verified by measuring the adsorption capacities of both samples annealed at both temperatures (100 and 450 °C). Figure 2.24 clearly shows that annealing decreases the adsorption capacity.

Of the two effects described above (sintering and defect creation by annealing), the sintering is more important with these materials, unlike on TiO_2, where the oxygen creation is more important.

2.7.4 Basalt, Siderite, Silica and Calcium Carbonate

In order to identify whether any of the rocks or minerals in the Sokolov clay exhibit oxygen exchange with CO_2 themselves or whether any other natural minerals exhibit oxygen exchange, we attempted to replicate the process with basalt (rock), siderite, silica and calcium carbonate (minerals). The natural basalt was obtained in the Císařský Quarry, Šluknov, Czech Republic, siderite in Rožňava, Slovak Republic, and calcium carbonate and silicon dioxide were purchased as powders from Sigma Aldrich. All the samples were dried in vacuum and annealed at 100 °C to remove any water and then exposed to $C^{18,18}O_2$. Silicon dioxide turned out to be inactive and after being exposed to $C^{18,18}O_2$ for one week, no $C^{18,16}O_2$ or $C^{16,16}O_2$ was detected. The other minerals—basalt, siderite and calcium carbonate—exhibited exchange activity. Figure 2.25 shows the conversion curves and the partial pressures of $C^{18,18}O_2$ and $C^{16,16}O_2$ throughout the experiment. Figure 2.26 shows the adsorption capacities of the respective samples.

Adsorption capacities of these minerals revealed that no trend in the adsorption capacity corresponds to the trend in half-lives of the exchange reaction. This proves that the oxygen atom mobility does not depend only on the particle size, surface area and oxygen defect presence on the mineral, but that it also depends on the chemical composition of the mineral. For example, calcite already contains (CO_3) which plays the role in the exchange as an intermediate. The geometry is therefore occupied by the (CO_3), and the mineral exhibits lower exchange activity (Civiš et al. 2015).

To sum up this chapter, CO_2 readily interacts with mineral surfaces commonly available in nature. It was found out that not only TiO_2, but other minerals exchange oxygen atoms with CO_2. Solid-gas interface is not a firm inert boundary, but a dynamic system where at room temperature, interaction takes place. These findings

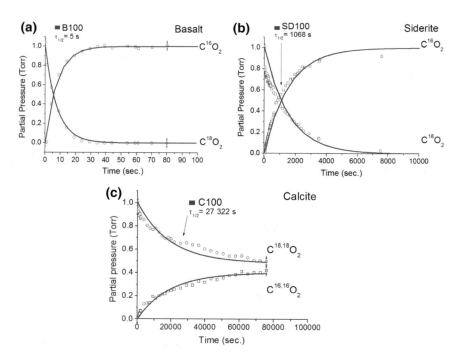

Fig. 2.25 Conversion curves and partial pressures for calcite, siderite and basalt. Basalt is the fastest of these samples in terms of the oxygen exchange while calcite, which already contains (CO_3) geometry, is the slowest. Reprinted (adapted) with permission from Civiš et al. (2016b). Copyright (2019) American Chemical Society

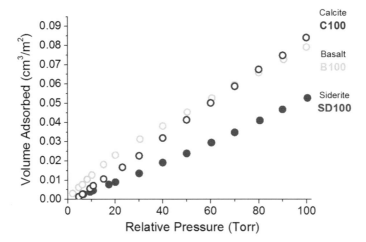

Fig. 2.26 CO_2 adsorption capacities of calcite, basalt and siderite annealed at 100 °C. Reprinted (adapted) with permission from Civiš et al. (2016b). Copyright (2019) American Chemical Society

should be used in the estimation of temperature on the early Earth through oxygen isotopic ratios, in planetary modelling and in the chemistry of interfaces.

Chapter 3
Photocatalytic Transformation of CO_2 to CH_4

Abstract The photocatalytic reduction of carbon dioxide is a process that has received significant attention in the literature of the second half of the twentieth century. This chapter shows how chemical conditions and the environment can envisage various aspects of this process and explains what consequences these findings may have on the origin of methane on terrestrial planets, which is shown on the case of Mars.

3.1 Methanogenesis

In the course of the experiments concerning the oxygen atom exchange between the solid and gas phases (Chap. 2), we noticed that UV radiation may alter the reaction route and direct its course towards CO or even CH_4. Pondering the source of hydrogen for methane, we first suspected water as there have been several studies already devoted to the matter [for a relatively recent review, please refer to (Habisreutinger et al. 2013)]. Its presence, however, was insufficient to explain the observed reactions. Also, annealing of the sample at 100 °C during its synthesis did not affect the photoreduction much. On the other hand, annealing at 450 °C slowed the photoreduction down. At last, it turned out that the missing agent was HCl, which was, as a leftover from the synthesis, present in the sample. Upon generalization, we show here that such a photoreduction is significantly affected by the presence of an acidic proton in the powdered samples of titania and other minerals. Again, using high-resolution infrared spectroscopy of CO_2, CO, HCl and CH_4 bands, it is possible to show the kinetic rate of the reaction, the effect of the proton and sketch a possible simplified reaction mechanism. It has also been estimated, as will be discussed later, that the photoreduction process is coupled to an oxygen transfer to Ti^{3+} centres in the catalyst. The final ratio has also been examined and it is shown below that the presence of HCl directs the reaction towards CH_4 while H_2SO_4, through a yet unknown or even unguessed mechanism, provides more CO than CH_4 in the reaction mixture.

The photoreduction, or 'methanogenesis', is a significant area of research which becomes more relevant with every day. The increasing concentration of greenhouse gases (especially CO_2) threatens to destroy the once mild conditions on the Earth.

S. Civiš et al., *The Chemistry of CO_2 and TiO_2*,
SpringerBriefs in Molecular Science,
https://doi.org/10.1007/978-3-030-24032-5_3

Even last year, 2017/2018, continental Europe experienced one of the mildest winters in recent history while North America saw snow, blizzards and severe cold. All over the world, temperature records both high and low have been broken that year more times than ever. The mitigation of global warming through carbon dioxide reduction is therefore one of the main tasks for humankind. There exist alternatives to CO_2 reduction as well, such as storage underground, but a photochemical conversion using sunlight offers an effective and cheap effective way of reducing CO_2 levels in the atmosphere while producing fuel in the form of methane. It is well known that photo-irradiated metal-oxide semiconductors serve as catalysts for the reduction providing a suite of molecules such as methane, carbon monoxide, formic acid, formaldehyde, acetylene or methanol (Inoue et al. 1979; Demont and Abanades 2015). The photochemical reaction pathway from CO_2 to methane involves several reaction steps producing both stable and unstable molecular intermediates (51) and has not yet been definitively made clear. Examining the pathway and the effects on the reaction through addition of acids is a part of this study. Its explanation would greatly help the research in the area of environmental technology and facilitate the construction of commercial-use photoreactors for the photocatalytic reduction.

3.2   The Methanogenesis Experiment

Many studies demonstrated that carbon dioxide can be converted to methane on TiO_2 surface by a photocatalytic process (see, e.g. (Habisreutinger et al. 2013) and references therein). Chemical conversion of carbon dioxide into energy containing fuel offers cheap and feasible solution. Such photocatalytic technologies towards carbon dioxide conversion into fuel have attracted the attention of many researchers and swiftly became promising in application. Known data about the reduction of CO_2 to methane, i.e. 'methanogenesis' on titanium dioxide (Dimitrijevic et al. 2011; Shkrob et al. 2012b) indicate the existence of photo-induced charge-transfer reactions involving two or more electrons from the titanium dioxide catalyst (Dimitrijevic et al. 2011, 2012). Another possibility is a one-electron stepwise reduction from the solvent bulk, which involves Ti^{3+} centres with a trapped electron on the mineral surface.

Studies have already been carried out by many, which address the importance of the Ti^{3+} centres during heteroatomic photo-induced reactions. For example, Zapol et al. (Shkrob et al. 2012a, b) describe two fundamental reactions on the Ti^{3+} centres:

$$Ti^{3+} + A^{ad} \rightarrow Ti^{4+} + A^{-\cdot} \tag{3.1}$$

$$Ti^{3+} + RB_{ad} \rightarrow Ti^{4+}B^- + R^{\cdot} \tag{3.2}$$

The reduced Ti^{3+} centres in this process serve as electron donors and the interacting molecules as electron acceptors. In the first reaction, the scheme shows only an electron transfer from one atom to another. In the second reaction, the authors present a different mechanism. This mechanism is limited to organic adsorbates, where R signifies an organic group and B a base, usually involving a heteroatom (O, N), for which the Ti^{3+} centre serves as an acceptor and thus a covalent bond is formed. This second mechanism has potential applications in sunlight-mediated conversion of CO_2 to methane or the chemistry of Mars and the Saturn's moon Titan.

In our photochemical experiments, three UV broadband light sources (each of them 300–400 nm, λ(max) = 366 nm, 160 W, E27 Omnilux Lamp) in air flow cooled photoreactor vessel have been used to irradiate powdered TiO_2 in the presence of 10 Torr of CO_2. At that stage of the experiments, three materials have been tested as photocatalysts:

(1) Crystalline TiO_2 anatase synthesized in an all-glass vacuum apparatus from $TiCl_4$ (99.98%, Sigma Aldrich, USA) by vapour hydrolysis with deionized ice. Detailed description of the synthesis has been described in our earlier research papers (Civiš et al. 2011, 2012, 2014, 2015; Kavan et al. 2011; Ferus et al. 2014a). This sample has been designated A100, where the number refers to annealing of the sample at 100 °C after the synthesis. Such sample contained 14 wt% of HCl.

(2) A part of the anatase A100 annealed at 450 °C for 6 h. This sample was redesignated A450. The annealing removes all H_2O and HCl traces from the sample. 10% H_2SO_4 has consequently been added to the sample.

(3) A450 sample without the added H_2SO_4.

After the irradiation, the products were monitored with a FTIR high-resolution spectrometer (Bruker GmbH, Germany) (Boháček et al. 1990; Civis et al. 1998; Ferus et al. 2008; Civiš et al. 2012, 2013). The detection and the irradiation were both conducted in a 200 ml T-shaped glass optical cell with CaF_2 windows for infrared analysis and a quartz finger for UV irradiation. The cell is depicted in Fig. 3.1. During the irradiation, the cell was placed in a cylindrical reactor externally cooled to approx. 50 °C. At selected time intervals, the cell was taken out and transferred to the spectrometer, where the actual composition of the gas phase was determined. The spectral measurements were performed in the range 2000–5000 cm^{-1} with the resolution 0.02 cm^{-1} using a KBr beamsplitter and a nitrogen-cooled InSb detector. The spectra were apodized with the Blackman-Harris 3-Term apodization function after the experiment.

The main products of the photocatalytic reduction were identified as CO and CH_4. Their formation was qualitatively observed in both A100 and A450 with acid experiments. Their formation, however, was not observed in the pure A450 sample. Figure 3.2 shows the most prominent rovibrational bands of the measurement. From the top in order, the spectra show, after 70 h of irradiation: A450 + H_2SO_4, A100, pure A450, methane standard, CO_2 standard, CO standard. Even though methane formation was not observed using the pure A450 sample, 0.05 Torr of CO was formed in this process.

Figure 3.3, on the other hand, shows a detail of the ν_3 methane rovibrational band and its evolution in the sample A100 during the irradiation.

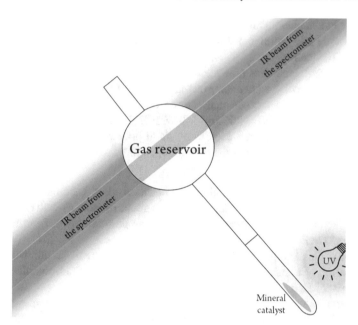

Fig. 3.1 Scheme of the optical cell equipped with a quartz finger for calcination and irradiation

As in the case of the oxygen isotope exchange experiments, the spectral lines were fitted with a Gaussian-Lorentzian function in the OPUS Software (Bruker GmbH, Germany), and their integrated intensities were fit through a calibration of standards assigned their actual corresponding pressures. The resulting partial pressures of each reactant were plotted and are shown in Fig. 3.4. In the A100 sample, after the irradiation, the CH_4 concentration is significantly larger than that of CO. In the case of the A450 anatase with H_2SO_4, the trend is reverted and CO dominates the product mixture.

Time evolution of their gas phase concentrations is depicted in Fig. 3.4. At the end of the irradiation, the concentration of CH_4 in A100 anatase sample is significantly larger than that of CO. On the other hand, A450 sample with H_2SO_4 produces distinctly more CO than CH_4 (the methane band is about 4 times weaker than that of CO and is therefore difficult to see in Fig. 3.2. Treatment of CO_2 in the presence of A100 anatase (panel B) resulted in formation of 0.33 Torr of methane, while A450 (panel A) mixed with 10% H_2SO_4 provided 0.26 Torr of CH_4. In case of A100 sample irradiation, the ratio of carbon monoxide to methane was about 80 times higher than in case of experiment with A450. In the case of the A450 + H_2SO_4 sample, the CO/CH_4 ratio was nearly 40, while in the A100 sample, the ratio was 0.25.

Mechanism of CH_4 and CO formation has been discussed in details by Zapol et al. (Shkrob et al. 2012b). A drastically simplified scheme is shown in Fig. 3.5. In the first step, CO_2 is adsorbed in the TiO_2 mineral surface. The UV irradiation excites an electron from the crystal structure and the electron migrates towards the adsorbed

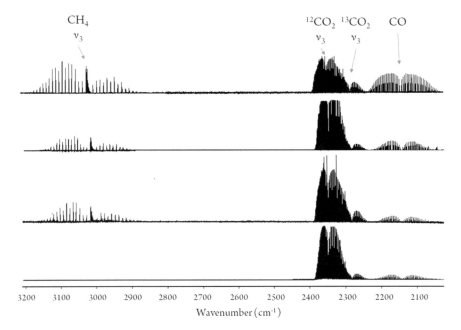

Fig. 3.2 Spectra of the three samples after 70 h of UV irradiation. The top spectrum shows measured standards of methane, CO_2 and CO. The second-from-top spectrum shows the sample A450 with H_2SO_4, the third-from-top spectrum the A100 anatase and the bottom spectrum the A450 pure sample (no acidic proton source available). Reprinted from Civiš et al. (2016a) Copyright (2019), with permission from Elsevier

molecule. Further, reaction of the adsorbed $CO_2\cdot$ with a proton leads through a cascade of further steps towards methane. Along the way, the reaction intermediates include, e.g. a formyl radical, HCO·, glyoxal, OHC–CHO and acetaldehyde (Ferus et al. 2009, 2011, 2014b). The final step is the photolysis of acetaldehyde and leads to the formation of CO and CH_4 in one step. This mechanism proposes the ratio of CO and CH_4 to be exactly 1. The proposed mechanism by Zapol et al. is therefore either wrong or there exist another side or reverse reactions which modify the observed ratios. Such side/reverse reaction has not yet been discovered.

In addition to methane and CO synthesis, we have also previously showed that the synthesis of acetylene is possible using a XeCl laser (Civiš et al. 2011).

3.3 The Case of Methane on Mars

Photocatalytic reduction of carbon dioxide has, besides environmental protection applications, relevance to the chemistry of Mars, especially in relation to its atmosphere. The constituents used in the studied process, CO_2, mineral catalysts, UV

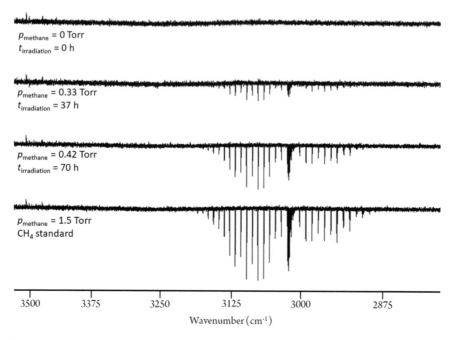

$p_{methane}$ = 0 Torr
$t_{irradiation}$ = 0 h

$p_{methane}$ = 0.33 Torr
$t_{irradiation}$ = 37 h

$p_{methane}$ = 0.42 Torr
$t_{irradiation}$ = 70 h

$p_{methane}$ = 1.5 Torr
CH_4 standard

3500   3375   3250   3125   3000   2875

Wavenumber (cm^{-1})

Fig. 3.3 Relative changes observed in the methane absorption band intensity in the experiments with the A100 sample. Reprinted from Civiš et al. (2016a) Copyright (2019), with permission from Elsevier

radiation and HCl are all present on Mars (Gordon and Sephton 2016). Also, the Martian soil contains a certain percentage or chlorate and perchlorate salts, as discovered by Hecht et al. (2009).

Here we present results of our experiments of CO_2 reduction to CO and CH_4 in Mars-like laboratory conditions. Apart from anatase (the same as the one used for isotope exchange experiments), we have tested the activity of montmorillonite, the Nakhla meteorite and other minerals. The final kinetic results for montmorillonite and anatase are shown in Figs. 3.6 and 3.7, respectively. Concentration of methane and CO was monitored in both samples and the time evolution of their production was plotted. The plots were then fitted with a first-order kinetic equation (Civiš et al. 2017). Overall, titanium dioxide shows a stronger tendency to form methane than montmorillonite. Montmorillonite shows significantly larger production of CO, which could be caused by either a different mechanism than on TiO_2 or by a yet unknown reverse process.

Martian surface is covered with regolith. Regolith is a fine-ground material which contains approx. 1% of TiO_2. In our experiments, we used the A100 sample described above and annealed it at 200 °C, which resulted in its redesignation as A200. The sample was bereft of water but retained 14% (w/v) of HCl embedded in its structure. About 0.5 g of the sample was annealed for 2 h and after that, the sample cell was filled with 6 Torr of CO_2. Such a sealed sample has been irradiated by UV and the

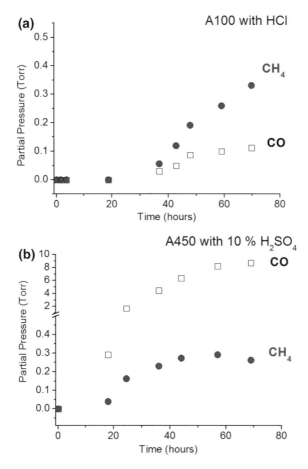

Fig. 3.4 Formation of main products in CO_2 treatment in the presence of A100 with HCl is depicted in the panel A and in the presence of A450 treated with 10% H_2SO_4 in the panel B. Partial pressure (Torr) of CH_4 is marked in blue circles and CO is marked in black hollow squares. Reprinted from Civiš et al. (2016a) Copyright (2019), with permission from Elsevier

composition of the gas phase was monitored using FTIR. This experimental set-up is fairly similar to that discussed above, but it has been fine-tuned in order to better represent actual Martian conditions. Its relevance is described in the following paragraphs.

The mean UV flux on Mars throughout the ages has been modelled by Ronto et al. (2003). Due to the evolution of the Solar System, the current UV flux on Mars is lower than in the past. It is still, however, about three orders of magnitude higher than on present-day Earth (Ronto et al. 2003; Schuerger et al. 2003) in the soft UV region. Also, Mars currently possesses a thin atmosphere dominated by CO_2. Various mineral catalysts together with some content of HCl are also present. These conditions are therefore well represented in our experiments with photocatalytic reduction.

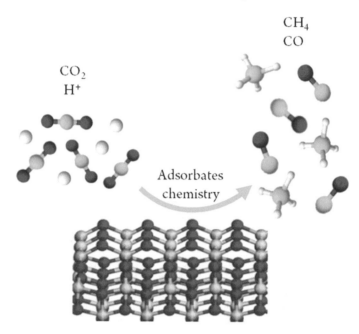

Fig. 3.5 A simplified scheme of the CO_2 reduction to CH_4. Carbon dioxide is first adsorbed on the TiO_2 surface upon interaction with a proton and an electron; the CO_2 molecule is activated and through series of reaction intermediates proceed towards carbon monoxide and methane. Water is formed as a side product of this reaction

The question of the seasonal variation and production rates of methane on Mars are to a certain extent presented in a paper by Civiš et al. (2017). These discussions are, at the current stage, pending the creation of a planetary model, which our laboratory is currently working on. The model has to include data from previous models of UV fluxes, atmospheric processes, dust swelling, mineralogy and elementary composition of the Martian surface as well as data obtained by NASA rovers currently exploring the planet Mars. It should be mentioned here that there are other explanations for the seasonal variation on Mars, most notably by Moores et al. (2019), who claim that the seasonal variation is consistent with regolith adsorption and diffusion.

Near-future outlooks and tasks in this field include testing of other materials effective for the photocatalytic reduction process, which may both be important in planetary chemistry of Mars and global warming on Earth, testing of the product mixture stability, determination of the mechanisms of the reaction cascade and possibly developing a more efficient set-up, which, in theory could be used in commercial applications and real-life technological solutions.

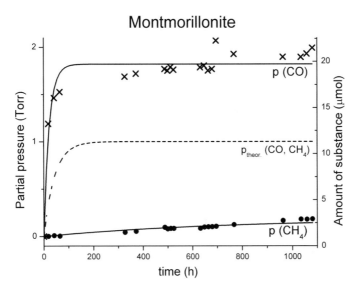

Fig. 3.6 Observed partial pressures of methane and CO in the experiment using montmorillonite. The dashed line represents a theoretical concentration of both CH_4 and CO were their kinetic rates equal as presumed from the proposed reaction mechanism. The image was originally published by Civiš et al. (2017)

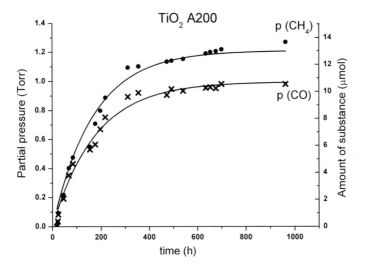

Fig. 3.7 Observed partial pressures of methane and CO in the experiment using TiO_2 anatase. The image was originally published by Civiš et al. (2017)

3.4 A Quest for the Methanogenesis Oxidation Products

The Martian surface is rich in both organic and inorganic compounds. Among the most famous are the perchlorates, whose origin has not been explained in a satisfactory way. Before we immerse ourselves in the experiment itself, which will show that perchlorates, chlorates, methyl chlorides, CO and methane share a common origin, we must pay a brief attention to the conditions on Mars. After that, we will explore the reaction, which is again the photocatalytic reduction of CO_2. This UV-initiated process on Mars begins with an atmosphere containing CO_2 and a source of H^+, in this case HCl, in contact with a catalytic surface. We will show that the oxidation products of this reaction are the (per)chlorates and methyl chlorides. In the latter sections, we will discuss the decomposition rates and the photochemical stability of the created molecules. This will lead to a prediction of a steady-state concentration of perchlorates in the Martian soil.

3.4.1 Chlorine-Bearing Species

The presence and origin of many of chlorine-bearing molecules on Mars remain to this day a mystery. The most famous of these molecules are the perchlorates, chlorates, chlorobenzene and chloroalkanes, such as methyl and ethyl chlorides (Glavin et al. 2013; Freissinet et al. 2015). An overview of the discussed molecules is shown in Fig. 3.8. For example, chlorobenzene is considered important, because any abiotic production route seems improbable (Freissinet et al. 2015). One possibility, which is widely discussed, is that this compound was created in the Curiosity analysis oven from a chlorine-containing molecule and the benzene ring in the course of the heating and does not come from the Martian soil after all. The second option is the existence of life on Mars.

The other important Martian species is perchlorate. Perchlorates were observed in the Martian soil by the Thermal and Evolved Gas Analyzer (TEGA) and the Microscopy, Electrochemistry and Conductivity Analyzer (MECA) instruments, which were part of the Phoenix Mars lander in 2008. The lander observed perchlorates at its landing site, the Green Valley, Vastitas Borealis, Mars, at concentrations up to 0.6 wt%, which is a significant content in the soil (Hecht et al. 2009). Their presence was later confirmed by the Curiosity rover's Sample Analysis on Mars laboratory (Glavin et al. 2013). The most common forms of perchlorates are magnesium perchlorate ($Mg(ClO_4)_2$) and calcium perchlorate ($Ca(ClO_4)_2$). In this state, the perchlorates are chemically relatively stable and can accumulate in the soil. The origin of perchlorates seems to be, due to their significant content in the soil, an important chemical pathway in the Martian chemical network. The origin of perchlorates was pondered by Smith et al. (2014), who created a chemical model in which they used reactions of perchlorate generation in the atmosphere similar to those of the Atacama Desert between the foothills of the Andes and the Chilean Coast Range. Due to its

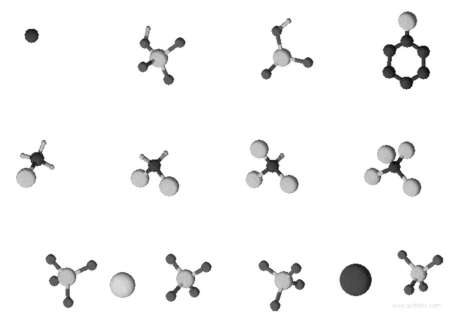

Fig. 3.8 Important molecules on Mars. Top row: methane, perchloric acid, chloric acid, chlorobenzene, mid row: methyl chloride, dimethyl chloride, trimethyl chloride, tetramethyl chloride, bottom row: magnesium perchlorate, calcium perchlorate. Atoms are coloured in the following manner: carbon—black, hydrogen—grey, oxygen—red, chlorine—green, magnesium—cyan, calcium—blue

setting in the plateau in between the two mountain ranges and therefore its relatively high altitude, moisture advection from both the Pacific Ocean and the Atlantic Ocean is prevented and the desert area lies in a two-sided rain shadow (Veblen et al. 2007). This causes the Atacama to be the driest desert on Earth. In a region ca 100 km south of the town Antofagasta, in a region which averages 3000 m elevation above the sea level, the soil of the desert has been compared to that of Mars. Interestingly, the region is used for filming Mars scenes in motion pictures. Most notably, however, in 2003, researchers from Mexico and the USA subjected the soil to chemical analysis and pointed out the similarities with the lifeless and organic material depleted Martian soil (Navarro-González et al. 2003). Since then, the research at Atacama is used as inspiration for the research on Mars. Smith et al. (2014) claim that according to their model, similar mechanism of perchlorate creation of Mars is likely. The authors evaluate the quantitative yields of perchlorates in their scheme and come to the conclusion that the obtained concentrations are too low to explain the measured abundance of perchlorates in the Martian soil. Since the model accounts for atmospheric chemistry reactions, the authors infer that heterogenous non-gas phase processes must supplement the production rate and that these must be explored in more detail in the future. A promising heterogenous non-gas phase scenario might be for instance the UV irradiation of titanium-containing crystals in aqueous solutions of chlorides (Schuttlefield et al. 2011). The scenario is nowadays problematic, because it infers the presence of liquid water on Mars. The lack of geological evidence for glaciation across the Martian surface suggests that even in the Noachian era Mars did not sustain a northern liquid ocean on the surface (Wordsworth 2016).

Perchloric acid itself is highly reactive oxidant. It can therefore form perchlorate salts, which can accumulate in the soil, where their content reaches up to 1% (Davila et al. 2013) (compare to 0.6 wt% measured by the Phoenix Mars lander). Experiments with the viability of *Bacillus subtilis* showed that the bacteriocidal effects of perchlorates are enhanced by iron oxides, peroxides and UV radiation (Wadsworth and Cockell 2017). The presence of perchlorates therefore speaks clearly against the survival of life on Mars (Brown 2008).

3.4.2 Methane on Mars

Methane is yet another intensely studies species on Mars. Its discovery ignited discussion about the presence of life on Mars in the first place (Lovelock 1965; Sagan et al. 1993). In terms of exoplanetary chemistry, methane has for long been considered a biosignature gas. Nowadays, when outgassing and atmospheric chemistry of exoplanetary bodies is better understood, the definition has been altered and now methane together with oxygen are considered as a single biosignature feature (Hu et al. 2012). The initial debate was inspired by the fact that 95% of the Earth's methane nowadays is of biogenic origin (Atreya et al. 2007). The assumption was that Mars could harbour hydrogen-utilizing bacteria, which reduce CO_2 to CH_4 by direct hydrogenation. Among the discussed abiotic sources of atmospheric methane on Mars were, e.g. releases from subsurface aquifers or wet deposition on the surface during winter and a subsequent release in summer (Oze and Sharma 2005), serpentinization of olivine (Oze and Sharma 2005), synthesis in a high-pressure and high-temperature hydrothermal fluid (Welhan 1988), volcanic outgassing (Geist 1992), cometary impact (Fries et al. 2016), meteorite impact, subsurface aquifers (Hu et al. 2016) or slow decomposition of subsurface organic matter reservoirs (Shkrob et al. 2010).

Methane was detected on Mars in 2003 by Mumma et al. (2009) and later by Krasnopolsky et al. (2004) and others (Formisano et al. 2004; Fonti and Marzo 2010; Geminale et al. 2011). These were all ground-based detections or satellite detections. In 2015, the Curiosity rover detected methane in situ on Mars. The rover landed in the Gale crater, where it now moves about the foothill of the Aeolis Mons. Chris Webster and the team from NASA JPL published a thorough analysis of measurements by the rover over the first two years (exactly 20 months) of the rover operation and concluded that the background concentration of methane on Mars was 0.69 ± 0.25 ppbv (Webster et al. 2015). The term background concentration was used, which means that data from local occasional plumes, such as were reported by Mumma et al. in 2003 were disregarded. This background, or steady-state, concentration was refined to 0.41 ± 0.16 ppbv in June 2018, when Webster et al. (2018) published data and analyses of the rover measurements over the five years of the rover activity on Mars. The authors also compared the seasonal variation trends to the seasonal variations of water, temperature and pressure, i.e. mostly CO_2. Although chemically stable in exoplanetary models of planetary atmospheres (Hu et al. 2012), methane is oxidized upon contact with the surface or with the help of the UV radiation and consequently exhibits a limited lifetime (Westall et al. 2015). Therefore, had no new sources of methane been present, it could not be detected with stable concentrations (Lippincott et al. 1967; Nair et al. 1994; Seager et al. 2016). This logically implies that the atmospheric methane must be replenished in some way. Photochemical models usually predict a lifetime of ~300 years (Nair et al. 1994). Chemical models predict lifetime in the order of 10^4 years. The seasonal variation, however, exhibits variation throughout the Martian year and the observed lifetime therefore lies in the order of months.

3.4.3 Planet Mars

Current Mars resembles a desert planet with no life in it and a very low organic compound content. The surface is covered with regolith, a fine particulate material with high content of iron oxides. These oxides cause the reddish hue of the Martian surface. The regolith also contains up to 1% of TiO_2, as has been stated above, and admixtures of clays. The upper crust of the planet is mostly basaltic, spotted with sites of pyroxene, olivine and plagioclase (Ehlmann and Edwards 2014). In places where the Noachian (>3.7 Ga) crust is exposed and clays are on the surface, as well as in Noachian and Hesperian (3.7–3.1 Ga) paleolake deposits of clays and sulphates, alteration through geologic processes suggests that the climate of Mars in the past was wetter and warmer. The alteration, however, is neither frequent nor thorough enough to support the idea of a sustained liquid water ocean, at least on the northern hemisphere. Sites with exposed Amazonian (present–3.0 Ga) clays exist as well but are even more scarce. Fries et al. (2016) also show that primordial Martian clays formed in a supercritical atmosphere or steam. Both primary and secondary alteration of rocks by water shows that up to 3.5 Ga, Mars preserved liquid water on its surface to a certain extent (Hu et al. 2015; Jakosky et al. 2015). Later, the water was gradually lost and newer findings of water activity are much sparser (Ehlmann and Edwards 2014).

This barren wasteland is immersed in about 6 mbar of CO_2 dominated atmosphere. The atmosphere consists mainly of CO_2 (95.32%), nitrogen (2.7%), argon (1.6%), oxygen (0.13%), carbon monoxide (0.08%). Apart from these, the atmosphere contains minor amounts of nitrogen oxides, water, krypton, ozone, xenon, methane, nitrogen oxides and neon. The composition and pressure of the atmosphere are kept at equilibrium with the polar caps, which are composed of frozen CO_2 and in times of winter on the respective hemisphere deplete the atmosphere of this gas. Also, the composition of the atmosphere is strongly influenced by the presence of the regolith, which adsorbs atmospheric gases and acts as an efficient catalyst in many a reaction.

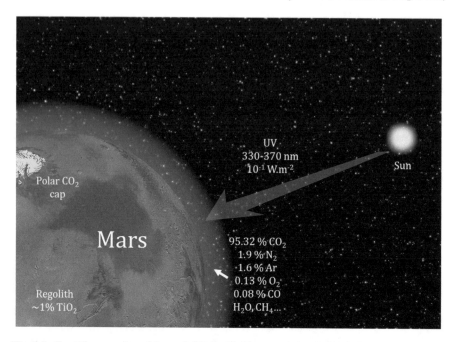

Fig. 3.9 Conditions on planet Mars suitable for the photocatalytic reduction of CO_2

The overall chemistry of the atmosphere is presumably powered by the incident UV radiation, because the atmosphere is thin and the UV penetrates to the surface. The actual surface UV flux is affected by the effective thickness of the atmosphere, the ozone concentration at the respective geographic location and dust swirling. The UV radiation flux on the surface is in the order of 10^{-1} W m^{-2} on the 330–370 region. Below 200 nm, the flux lowers to 10^{-4}–10^{-2} W m^{-2}. The UV radiation and the regolith create an ideal site for the photochemistry to proceed (Shkrob et al. 2010; Civiš et al. 2017), including, possibly, the photocatalytic reduction of CO_2 to methane. The conditions are shown in Fig. 3.9.

3.4.4 Laboratory Experiments

The actual laboratory experiment was performed by sealing together in a specially developed ampule 6 Torr of CO_2, 6 Torr of HCl, a catalyst (TiO_2, montmorillonite, the Nakhla meteorite and others) and this mixture was irradiated with a Hg broadband UV lamp with peak intensity at 350 nm and the total power 160 W. The glass ampoule was developed specifically for this purpose and is shown in Fig. 3.10.

The lower part of the ampoule is made of quartz glass, which is transparent to UV radiation. In this part, the photoactive catalyst is stored and the actual reaction takes

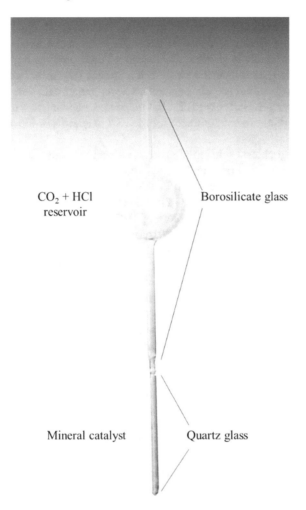

CO$_2$ + HCl
reservoir

Borosilicate glass

Mineral catalyst

Quartz glass

Fig. 3.10 Sample ampoule developed for the purpose of the photocatalytic reduction of CO$_2$. Reprinted (adapted) with permission from Civiš et al. (2019). Copyright (2019) American Chemical Society

place. This quartz tube is connected to a glass reservoir, which contains the reactants (CO$_2$, HCl). The standard borosilicate glass is transparent to UV radiation in the desired region between 2200 and 3500 cm^{-1}. The sample gas phase was monitored by IR spectroscopy at regular intervals. In the course of our work, we found out that an excess of protons (H$^+$) in the sample positively enhances the reaction rate. So far, HCl and H$_2$SO$_4$ were tested. In the case of Mars, HCl is probably the more interesting acid. The acid itself has never been detected directly on Mars, but its upper limit was estimated by Hartogh et al. (2010). However, various studies show that HCl is present on Mars in various forms and phases (Wong et al. 2003a; Catling et al. 2010; Ehlmann and Edwards 2014; Smith et al. 2014). The said works describe HCl creation in various processes on the surface and in the atmosphere. None of these was experimentally proven yet, but all of them carry a logic behind it, so that it seems likely that their discovery is a matter of the correct probe or piece of equipment, which would search for them on Mars, rather than their non-existence. The high reactivity of HCl should also be considered, meaning that it is possible that the production rates are not sufficient for the HCl to attain a stable concentration in the gas phase.

We used hydrochloric acid simply because it well explains the laboratory formation of perchlorates, chlorates and methyl chlorides, all of which were detected by the Curiosity rover (Freissinet et al. 2015). The methanogenesis itself is not limited to that, though, and any source of protons is feasible for the reaction. Previous experiments have shown that an excess of H_2O (l) hinders the reaction under our experimental conditions (Civiš et al. 2016a). For this reason, the sample was filled with HCl vapour rather than HCl solution.

The main idea behind this experiment, however, was not only the testing of the methanogenesis process, but also the search for the oxidized products of the reaction. Chemical laws of equilibria dictate that charge in a closed system must be conserved. Since our system is closed from the electric point of view and since CO_2 is reduced to methane, something must inevitably be oxidized in the sample. Considering the contents of our sample, the species available for oxidation are water and the chlorine atom. These together may yield O_2, O_3, Cl_2 and chlorine oxides and oxyanions. If the oxidation products were O_2 and O_3, the question would still remain: What of the chlorine? Therefore, we opted to search for chlorine-containing species first.

After a gas phase, steady state was reached and the reaction proceeded no further (HCl was depleted), we opened the sealed samples in an Ar atmosphere and added 500 μL of 10% degassed, CO_3^{2-} free KOH. If perchloric and chloric acids were present, as was assumed, they would react to form $KClO_4$ and $KClO_3$. The sample was then analysed by X-Ray photoelectron spectroscopy (XPS). This analysis indeed confirmed the presence of perchlorates and chlorates in the sample. Figure 3.11 presents a detailed spectrum of the samples in the Cl 2p region (left) along with standard reference measurements (right). The measured spectra (dots) were fitted to Cl $2p_{1/2}$ and $2p_{3/2}$ chlorine emission lines. It is clear from the figure that perchlorates, chlorates and chlorides were all detected in the sample, while perchlorates were the main product. The chloride anion is present in the anatase sample only. This may indicate either a reduction of chlorates in ambient conditions or a residual chloride from the sample, which in the case of anatase contained hydrochloric acid in its structure rather than on its surface.

The presence and creation of perchlorates were considered by Carrier and Kounaves (2015). Their experiments, which were similar to ours in many ways, showed the presence of perchlorates and chlorates. The authors come to the conclusion that 'perchlorate formation is most likely an ongoing process and is occurring globally on a continuous basis wherever chloride-bearing mineral phases exist'. Unfortunately for the authors, however, they did not analyse the gas phase of their samples, for it is likely that they would discover methane as well. The fact, how-

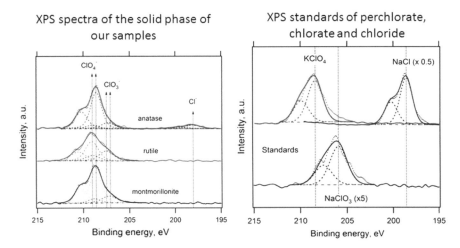

Fig. 3.11 Detailed spectrum of the samples in the Cl 2p region (left) along with standard reference measurements (right). The intensity scales are identical for all plots. The signals of ClO_4^-, ClO_3^- and Cl^- are labelled by arrows. Reprinted (adapted) with permission from Civiš et al. (2019). Copyright (2019) American Chemical Society

ever, that they independently reached the same conclusion concerning perchlorates as we did strongly supports the theory that perchlorates on Mars originate in the photochemistry of CO_2.

3.4.5 Methyl Chlorides

Along with the detection of other compounds, chlorohydrocarbons, namely chloroalkanes, were discovered on Mars. The Viking lander observed CH_3Cl and CH_2Cl_2 upon heating of the regolith (Biemann et al. 1977). The pyrolysis analysis of rock samples from Sheepbed mudstone (Freissinet et al. 2015) and Rocknest (Glavin et al. 2013) by Curiosity brought evidence of the presence of CH_3Cl. As of yet, their presence was mainly attributed to contamination or an unidentified reaction during the analysis process, which involves heating of the sample. Another discussed possibility is a reaction between Martian chlorine and terrestrial carbon from the rover. We observed the formation of perchlorates in our experiments. They were detected by selected ion flow tube—mass spectrometry (SIFT-MS) and gas chromatography—mass spectrometry (GC-MS) (Knížek et al. 2017) in 0.01–1% concentration. They were not detected by infrared spectroscopy, but this may be because either they were adsorbed on the surface or their concentration was too low for detection. The sample ampoule has a diameter and therefore the optical path of about 5 cm, which results in detection limits around 0.02 Torr of CO_2 or HCl in the sample, both of which are very strong absorbers in the infrared region.

This finding shows that methyl chlorides are created during the photoreduction. The mechanism of their creation is unknown, but presumably, chlorine species may react with the freshly created methane in the sample, which results in the production of the methyl chlorides.

3.4.6   The Nakhla Meteorite

The strengthen to plausibility of the photocatalytic reduction on Mars, we performed experiments on the Nakhla meteorite. The Nakhla meteorite fell to Earth on 28 June 1911 and was named after the nearby village, El Nakhla, Egypt. The meteorite is a Nakhlite type of the SNC group of Mars meteorites. These meteorites were ejected from the planet during its collision with a larger body. The piece of rock is ejected from the home planet and travels across the Solar System for an unknown period until it penetrates the Earth's atmosphere. A splinter of the meteorite was ground to dust powder and used as catalyst in our experiment. The gas phase spectra of the sample are shown in Fig. 3.12. The figure shows the detection of methane created on the Nakhla meteorite. The amount of available meteorite was very low (100 mg) and its activity is not high compared to the anatase sample. Therefore, the spectrum was zoomed 150× for clarity. Upon magnification, however, the presence of methane is without doubt.

Other minerals were tested for their activity in the photoreduction process as well. From the most efficient, in decreasing order, they are Al_2O_3, kaolinite, MgO an acidic and ferrous synthetic clays. They are natural minerals (apart from the synthetic clays, obviously) and therefore show that if the conditions are right, the methanogenesis may take place anywhere in nature and is not limited to laboratory or Mars. As for Mars, the photocatalytic reduction was also explored at $-196\,°C$ (liquid nitrogen cooling). Therefore, the reaction works at such a low temperature as well. This shows that on Mars, where temperatures reach down to $-120\,°C$, the reaction may proceed regardless. It is important to note, however, that with such a drastically different temperature, the reaction rates will probably differ and quantitative comparison of laboratory measurements and Mars should be approached with utmost caution.

A final mention should be made towards the seasonal variation of methane on Mars. Webster et al. (2018) describe seasonal variation of methane on Mars. They compare the variation to the behaviour of other species, such as CO_2 and water in the Martian atmosphere throughout the year and conclude that no definitive similarity between any two species can be observed. We published in 2017 an article (Civiš et al. 2017), where we propose and predict the seasonal variation of methane on Mars due to the photocatalytic reduction of CO_2. In this prediction, the content of methane, CO_2 and water does not fluctuate together, of course, because they are reactants and a product. It seems to us that the variation observed by Curiosity and our prediction may fit well together, but since the Curiosity experiment was not designed to evaluate this experiment, we can now not say what its results really signify. Also, before such prediction is either confirmed or disproved, a planetary model must be created, which

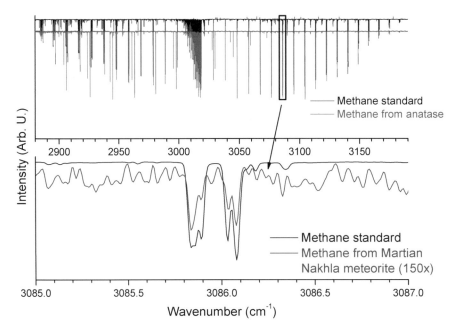

Fig. 3.12 Photocatalytic reduction of CO_2 on the Nakhla meteorite. The upper spectrum shows a methane band at 3085.7 cm^{-1} measured in standard methane gas at 0.69 Torr (black) and in gas phase methanogenesis experiment with anatase (red) after 1000 h of UV irradiation. The bottom spectrum shows the same line in the gas phase of the methanogenesis experiment with the Nakhla meteorite sample (red) upon UV irradiation for 72 h with a 160 W broadband Hg lamp (peak intensity at 350 nm). Reprinted (adapted) with permission from Civiš et al. (2019). Copyright (2019) American Chemical Society

would describe the magnitude of many interfering effects on a global planetary scale. Such model is being created now, but it will take time, effort and more data to make this model sensibly accurate. It should be noted that Moores et al. (2019) proposed a different mechanism of the generation of the seasonal variation through adsorption and desorption on the regolith.

3.4.7 Global Yield Estimates

The following section was originally prepared by Dr. Paul B. Rimmer, who is a planetologist at Cavendish Astrophysics Laboratory at the University of Cambridge for our mutual publication (Civiš et al. 2019). This section is reprinted (adapted) with permission from Civiš et al. (2019). Copyright (2019) American Chemical Society. We do not claim to be planetologists ourselves, but the section is necessary for the understanding of the continuity of the story in this book.

The effects described above need not be dramatic in order to explain the observed perchlorates on Mars and in order to make firm predictions about the present Martian surface. The surface methanogenesis we propose is sufficient even with trace amounts of hydrochloric acid, consistent with the sub-ppt concentrations predicted by Smith et al. (2014) We can predict the perchlorate content of Mars from this methanogenesis, under the assumption that the perchlorates are stable over Ga timescales and that this heterogeneous process is the dominant mechanism by which perchlorate salts are formed.

The methane concentrations observed to emanate from the Gale crater fluctuate seasonally, with a minimum abundance of 0.69 ppbv (Webster et al. 2015). At the same time, there is expected to be 0.1 ppbv of O and 0.1 pptv of OH near the surface (Hu et al. 2012). UV photons do not penetrate deeply enough to remove the CH_4 (Wong et al. 2003b), so we consider only destruction by O: and ·OH, which should dominate. The reactions for destruction are:

$$CH_4 + \cdot OH \rightarrow CH_3 + H_2O, \quad k_1, \tag{3.3}$$

$$CH_4 + O(^1D) \rightarrow CH_3 + \cdot OH, \quad k_2, \tag{3.4}$$

$$CH_4 + O: \rightarrow CH_3 + H_2O, \quad k_3. \tag{3.5}$$

Here, the rates constants are (from (Atkinson et al. 1992; Corchado et al. 1998; Bravo-Perez et al. 2005), respectively):

$$k_1 = 5.2 \times 10^{-13} e^{\frac{-1640K}{T}}, \tag{3.6}$$

$$k_2 = 1.4 \times 10^{-10}, \tag{3.7}$$

$$k_3 = 5.6 \times 10^{-10} e^{\frac{-6220K}{T}}. \tag{3.8}$$

The influx of the CH_4, Φ_{CH_4}[cm^{-2} s^{-1}] can be found by balancing the kinetics equation at steady state:

$$\Phi_{CH_4} = H_0 \left(k_1[OH] + k_2 \left[O(^1D) \right] + k_3[O] \right) [CH_4], \tag{3.9}$$

where $H_0 = 11$ km is the Martian atmospheric scale height. The left side of the equation represents the production of methane and the right side its destruction. This results in a surface flux of about $\Phi_{CH_4} = 1000$ cm^{-2}s^{-1}.

Now consider the overall reaction:

$$14\ CO_2 + 10\ HCl + 14\ H_2O + h\nu \rightarrow 7\ CH_4 + 7\ CO + 5\ HClO_4 + 5\ HClO_3 \tag{3.10}$$

Stipulating that this reaction alone explains the methane outgassing, that this minimum outgassing rate is the average rate over the entire Martian surface, and that perchlorates are stable over the lifetime of Mars [if sustained in solid state (Martucci 2012)], we can calculate how much perchlorate will be produced by this scheme.

If we think that at the end of the day the 5 $HClO_4$ and 5 $HClO_3$ will end up as 10 ClO_4^- in the form of some salt or other that means we expect ten molecules of ClO_4^- for every seven molecules of CH_4. Over the entire irradiated surface of Mars (half the planetary surface), about 7.0×10^{20} molecules of CH_4 are being released per second, meaning that about 1×10^{21} molecules of ClO_4^- are being generated every second. Over the past 10^9 years, 3×10^{37} molecules of ClO_4^- would have been generated.

Perchlorate (ClO_4^-) salts make up on the order of 0.5% of the mass of a rock at the surface of Mars (Navarro-Gonzalez et al. 2010). The density of these Martian rocks is 1.5 g cm^{-3} on average, and ClO_4^- molar mass is about 99.4 g mol^{-1}.

The depth of the observed perchlorate content, d, can be calculated by the following equation:

$$d = \frac{NM}{N_A w \rho S}, \tag{3.11}$$

where N is the total of produced molecules of ClO_4^- (3×10^{37} over the past 1 Ga), M is the molar mass of ClO_4^- (99.4 g mol^{-1}), N_A is the Avogadro constant, w is the mass fraction of perchlorates in the Martian soil (0.005), ρ is the average density of Martian surface rocks (1.5 g cm^{-3}) and S is the surface area of Mars (1.44×10^{18} cm^2). This rate of photochemical CH_4 production over time would therefore result in rocks with the observed perchlorate content 1 cm deep per Ga. If the amount of methane released was greater in the past, closer to the peak amounts of 7.2 ppbv observed (Webster et al. 2015), then the amount increases to 10 cm per Ga.

This process of photochemical methanogenesis explains the content of perchlorates accumulated during the existence of Mars, predicts that it extends between 5 and 50 cm into the Martian surface, depending on the past outgassing rates of methane, and also predicts an equal amount of carbon monoxide outgassed, which should be observable and coexistent with the outgassed methane. The observation of CO is essential for distinguishing this methanogenesis mechanism from biotic methanogenesis, as biotic methanogenesis will tend to produce CH_4 without the CO intermediate, both according to the reaction we consider and according to fundamental principles of redox chemistry and disequilibrium processes. The amount of CO should maybe differ from CH_4, since reverse reactions of CH_4 oxidation to CO may be present, but nonetheless, its presence and concentration fluctuations point to the abioticity of the methane source. Temporal variations of CO abundance on Mars have been already presented in the literature. The variations were, however, detected during one day on Mars (Bar-Nun and Dimitrov 2006) or in 11-year period and correlated to solar activity cycles (Krasnopolsky 2010). Seasonal variation has also been observed (Encrenaz et al. 2006), but with up to 50% uncertainty. These variations were also correlated to latitudinal variation (Krasnopolsky 2007). We are

not aware of any study which would systematically compare seasonal variation of
CO and methane [an explanation for the methane seasonal variation was proposed
by us (Civiš et al. 2017), by Webster et al. (2018) in June 2018 and by Moores et al.
(2019)].

3.4.8   The Reaction

An overall reaction figure was created and is shown in Fig. 3.13. The scheme shows
the reaction products, which have so far been detected. The products are separated
into groups according to the detection method used for their discovery. It should be
noted that the scheme does not portray a single reaction, but a reaction sequence takes
place in between the reactants and the products. On the whole, the reaction begins
with CO$_2$ and HCl. In the presence of a catalyst and UV radiation, the reactants
are consequently transformed to CO, CH$_4$ (detected by IR), CH$_x$Xl$_y$ (detected by
GC-MS) and ClO$_4^-$ and ClO$_3^-$ (detected by XPS). The perchlorate and chlorate are
volatile when in form of their acids and therefore KOH was added to the product
mixture after gas phase analysis. The KOH reacts with the acids and forms non-
volatile potassium salts, which can be detected by XPS.

The presented and discussed results are not limited to Mars or early Earth. When
the conditions are met, these processes may take place on any extraterrestrial body.
This realization is important for astronomers and astrochemists concerned with bodies in our Solar System, such as the Saturn's moon Titan or even exoplanets, whose
atmospheric compositions are currently on the verge of discovery.

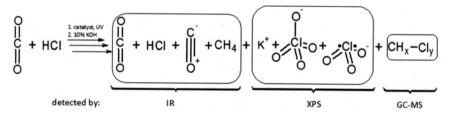

Fig. 3.13 Overall reaction scheme. The reaction begins with CO$_2$ and HCl. Then, in the presence
of a catalyst and UV radiation, the reactants are consequently transformed to CO, CH$_4$ (detected by
IR), CH$_x$Xl$_y$ (detected by GC-MS) and ClO$_4^-$ and ClO$_3^-$ (detected by XPS). Reprinted (adapted)
with permission from Civiš et al. (2019). Copyright (2019) American Chemical Society

3.5 Methanogenesis on Earth and Other Planets

The process of methanogenesis as shown in the case of Mars is not limited to Mars only. From the chemical point of view, any environment in suitable conditions may be subject to this transformation.

Nowadays, the ozone layer prevents most of the UV radiation from reaching the planetary surface. The mean spectral irradiance of the Earth's surface is about 10^{-1} W m^{-2} in the soft UV region (McKinlay and Diffey 1987). This is comparable to Mars. Ozone is transparent in this region (Hearn 1961) and the only notable difference in the flux between Earth and Mars is the distance from the Sun. The flux falls down with the inverse square law. The Earth is approximately 1 AU, or 150,000 km distant from the Sun. Mars is about 1.5 AU away from the Sun, or 228,000 km. The flux on Mars should therefore be roughly 2 times lower than on Earth, but of the same order of magnitude. Earth also has much denser atmosphere which attenuates the flux by the presence of admixtures or scattering. In the region below 300 nm, the flux falls rapidly due to the effect of the ozone layer. Below 290 nm (2006), the irradiance flux is less than 10^{-3} W m^{-2}.

Nevertheless, it is shown in the previous chapters that soft UV radiation, that is 300-400 nm, is sufficient to power the methanogenesis process on TiO_2. It might theoretically be possible to observe the methanogenesis on Earth nowadays. It is an abiotic way in which the atmosphere is depleted of CO_2 and enriched with CH_4, which is a process countering the effects of CO_2 emissions into the atmosphere. However, finding a place where TiO_2 is exposed to the atmosphere is a challenging task. Of the possible places, one would still have to select those that are at least wet (the presence of water) or better slightly acidic. Also, the atmosphere contains about 410 ppm of CO_2. Other molecules, such as O_2, which constitutes about 21% of the atmosphere, may be adsorbed on the catalyst as well, thus blocking the adsorption sites of CO_2. It would also be difficult to find pure TiO_2. In such a complex system with impure catalyst and reactants in mixtures and low concentrations, interfering reactions are almost a surety. All in all, with the methanogenesis having a slow rate, the overall effect on atmospheric composition on the current Earth would be negligible. On the early Earth, however, the situation might have been different.

3.6 Emergence of Organic Molecules on Earth and Elsewhere

As a final result stemming from our experiments, we noticed in the previous experiments that if the resulting mixture (CH_4, CO) is supplemented with traces of water and molecular nitrogen (common gases in planetary atmospheres) and exposed to high-energy density conditions, such as laser pulses ($\sim 10^2$ J) or electric discharge, complex organic compounds, such as nucleic acid bases are created (Civiš et al. 2017).

Photochemical models of the Sun's activity and Earth's composition suggest that the early Earth experienced less effective UV radiation. The overall UV power of the Sun was higher (Cockell 1998), but its dissipation and absorption were also more powerful, so the effective energy available as an energy source for the initiation of prebiotic reactions was less important than other energy sources, such as asteroid impacts into the early Earth's atmosphere (Chyba and Sagan 1992). The photoactive minerals, such as clays or titanium dioxide, were present on the early Earth as well (Ferris et al. 1996). Since all the necessary conditions—UV radiation, minerals, CO_2 in the atmosphere—were present on Earth, it can be assumed that the atmosphere was at least partly or temporarily converted from a CO_2-rich (neutral) to CO and CH_4-containing (partly reduced) atmosphere. These minor conversions could be on scales such as would not be seen in geological records. Also, it turns out from a recent zircon analysis, that the Hadean continental crust was more reduced than the modern crust (Yang et al. 2014). Moreover, besides CH_4 produced by methanogenesis, other sources, such as volcanic activity, granitic magmatism and outgassing, may have supplemented the methane stock on the early Earth.

An easy scenario may then be contemplated. The early Earth has either neutral or reduced atmosphere. If reduced, all is ready for the next step, which is prebiotic synthesis. If neutral, it may be at least partly reduced by the methanogenesis process. Actually, the initial state of the atmosphere matters in terms of the reaction kinetics, not plausibility. Next, the atmosphere containing CH_4 and CO, together with N_2, may be used for the creation of organic molecules, such as RNA bases (Ferus et al. 2014b). For this synthesis, a relevant energy source is required. One such source may be an extraterrestrial body impact. A passing mention should be made, that the first evidence of life on Earth coincides with the Late Heavy Bombardment era on Earth, some 4.1–3.8 Gya. It may be interpreted as a pure coincidence or as a supporting argument for the fact that extraterrestrial body impacts played a vital role in the creation of life rather than destruction of everything that was on Earth back then. It is worth mentioning that recently, opinions have arisen that the formation of the Solar System was calmer than previously thought and that the LHB may not have occurred after all (Christodoulou and Kazanas 2019).

An extraterrestrial body impact brings high-energy-density conditions, shock waves, shock rise in temperature up to 4500 K, hard UV radiation and creates a dense plasma. Experimentally, these conditions can be reproduced by a high-energy laser, if the beam is focused on the surface of the sample inside a closed sample cell. The interaction chamber from the Prague Asterix Laser facility is shown in Fig. 3.14. These impacts share a common feature, which is the presence of ·CN, ·NH, ·CH and vibrationally excited CO in the plasma. During the afterglow when the plasma cools down, the unstable species react and recombine. The main product is HCN. This was proposed by Sagan and Chyba (Chyba and Sagan 1992) and experimentally verified and further explored by us, as shown in Fig. 3.15. During impacts into such an atmosphere, nucleobases may be directly formed. If the energy is insufficient, further impacts into the HCN rich atmosphere lead to the nucleobase synthesis, as has been proven in many experimental works, e.g. (Powner et al. 2009; Sutherland 2016; Ferus et al. 2017a). The experiment in our laboratory was carried out using a

Fig. 3.14 Prague Asterix Laser generates a laser-induced dielectric breakdown in the sample. This hall laser is used for the reproduction of asteroid bombardment

large hall laser PALS (Prague Asterix Laser System). In the resulting mixture, we discovered that from CH_4, CO, N_2 and a mineral catalyst, nucleic acid bases, as well as acetylene, NH_3 and HCN, were formed.

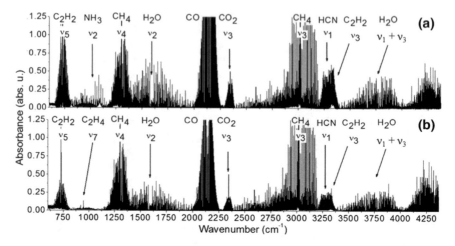

Fig. 3.15 Shock wave induced transformation of a reduced planetary atmosphere. Gas phase spectra of montmorillonite + HCl (**a**) and TiO_2 A200 (**b**), both in the presence of CO_2 + CO + CH_4 + N_2, transformed by the PALS laser. Main products of this transformation are HCN, acetylene, ammonia and ethane. Reprinted (adapted) with permission from Civiš et al. (2017). Copyright (2019) American Chemical Society

3.7 The CO_2 Cycle

In the past chapters, we saw that CO_2 undergoes a certain cycle different to the natural cycle on Earth.

We saw that CO_2 interacts with surfaces on a regular basis. The molecule can be adsorbed on a mineral surface and upon UV irradiation may be transformed into a molecule with carbon in a different oxidation state. This molecule is in our case methane or CO. This process is called photocatalytic reduction of CO_2 and is powered by soft ultraviolet radiation. On Earth, the source of UV is the Sun and, in the past, prior to the formation of the ozone layer, the process could have been more important for the overall atmospheric chemistry than it is today. The issue of the neutral or oxidized state of the atmosphere in the course of the chemical origin of life is therefore less important, because the CO_2-containing atmosphere could have been, at least temporarily and locally converted to a more reduced and prebiotic synthesis-favourable atmosphere. In fact, it is improbable that the atmosphere would have been converted whole.

On planets with CO_2 and without the ozone layer, the process must inevitably take place as well. This is the case of Mars, for example. There the concentration of CO_2 is low and the source of hydrogen atoms is weak, so the overall chemistry is not affected significantly. The same process can be found on exoplanets adjacent to different stars of various spectral types and composition. Overall, the methanogenesis may be in the appropriate conditions powerful enough to change the overall state of the atmosphere of any given planet. CO_2 is therefore converted into CH_4.

High-energy chemistry, thermochemistry or photochemistry that lead to the synthesis of prebiotically important molecules, such as amino acids or nucleic acid bases is more effective when the starting composition of the ambient atmosphere is rather reduced. This was first shown by Miller and Urey and the work of others has since verified this finding many times. The transformed atmosphere of CO$_2$, now predominantly CH$_4$ or CO, is therefore and ideal candidate for prebiotic synthesis, such as could have taken place on the early Earth. It would be bold to claim that the source of carbon/methane for prebiotic synthesis on the early Earth comes from the methanogenesis, but the possibility is that this process contributed to the overall synthetic process. Prebiotic synthesis of biologically important molecules from what initially was CO$_2$ is therefore a viable option.

The Earth experienced a violent youth. Initially, the planet was very hot and unstable. As it slowly cooled down, the conditions became more and more stable. This calming down was interrupted several times in the Earth's history by events such as the impact of Theia, the moon-forming asteroid, the Late Heavy Bombardment or the radioactivity. The Earth harboured many radioactive atoms that are nowadays already transformed and extinct. The radioactivity must have heated the surface of the planet significantly and surface radioactivity must have blocked the synthesis in the initial stages of the Earth's evolution. Only after the radioactivity levels decreased enough could life be formed. Volcanic activity was also significantly larger on the planet. These effects combined destroy any organic molecules present. Impact of an asteroid produces hot and dense plasma at the impact site and heats the immediate surroundings by several hundred K. Volcanoes spit out lava and radioactivity heats up the surface. In such violent conditions, burning is a common process. Organic molecules were therefore burned and destroyed. Since CO$_2$ is the stable oxidation product of carbon burning, the burning would release CO$_2$ again.

In this way, the cycle is complete. CO$_2$ undergoes transformation to a more reduced form of carbon, then takes place in prebiotic organic synthesis and in the end is recreated by some cataclysmic event. A graphical depiction of the cycle is shown in Fig. 3.16. The cycle is split into six sections/panels, which depict (in order) the oxygen atom exchange, photocatalytic reduction of CO$_2$, the connection to Mars, high-energy-event-driven synthesis of biologically active molecules, their polymerization to DNA and finally the destruction back to CO$_2$.

Fig. 3.16 A graphical depiction of the CO_2 cycle. In panel 1, the cycle shows the exceptional features of CO_2 which are, in this case, the oxygen atom exchange between CO_2 and the solid phase. Panel 2 shows the photocatalytic reduction of CO_2 to methane and CO. Panel 3 shows the connection to the chemistry of Mars and the origin of methane and perchlorates there. Human exploration attempts are shown through the presence of the Curiosity rover. Panel 4 is split in two. The upper part shows the transformation of a reduced, or partly reduced, planetary atmosphere to biologically relevant molecules through asteroid impact-driven chemistry. The lower part shows the PALS laser facility in Prague, where these impact-induced experiments can be replicated. Panel 5 then shows DNA, an omnipresent molecule in organisms, created through the polymerization of nucleic acid bases. Finally, panel 6 shows the destruction of these biomolecules through lightning or high-energy chemistry. The destruction of biomolecules produces again CO_2, which can start the cycle again

Chapter 4
Additional Views on Prebiotic Molecules

Abstract The field of the origin of life research is a relatively new but rapidly growing area of research. It is not the purpose of this book to review all scientific progress in this branch. Indeed, such book would have to be quite thicker, not to mention expensive. However, several ideas that have recently come up concerning the early Earth and life's origins include CO_2 and/or TiO_2 as the main protagonists. We wish to mention to a certain extent these ideas and show that photochemistry might have played a vital role in the process by which we came to being.

4.1 Primordial Earth Scenarios

Planet Earth arose from the protoplanetary disc ca 4.6 billion years ago. The planet, as well as other rocky planets in the Solar System, is formed by accretion of material from the disc around the Sun. There are two distinct ways of planetary accretion—hot and cold. Cold accretion is slower (100 million—1 billion years) and allows the planet to form by gravity-driven aggregation of solid bodies in the Solar System. Gravity of the growing body and radioactive elements, such as ^{235}U, ^{238}U, ^{234}Th or ^{40}K eventually heat up the planet, which melts at about 1500 K. At this point, a so-called iron catastrophe occurs, where iron and heavier elements fall to the centre of mass of the body and lighter compounds migrate to the surface in violent outbursts. In this way, the planet is differentiated and later cools down.

The hot accretion is a faster process (100,000 to 10 million years). The planet in this scenario forms from already molten material and rather rapidly. The rapid accretion does not allow for sufficient heat transfer and radiation and the planet accretes molten. The differentiation and stratification are therefore continuous. The planet eventually cools down and forms a stable crust and, in the case of the Earth, maintains a molten Fe–Ni core. Nowadays, the latter scenario seems to be the correct one (Shaw 2006; Beatty et al. 2015). It is believed that the Earth experienced an increased period of asteroid bombardment about 4.1–3.8 Ga, which is known as the period of the Late Heavy Bombardment. Roughly 3.8 Gyr ago, the period of the Late Heavy Bombardment ended. This was, as the name suggests, a period when the Earth was intensely bombarded by the surrounding material due to the ongoing stabilization

© The Author(s), under exclusive license to Springer Nature Switzerland AG 2019   69
S. Civiš et al., *The Chemistry of CO2 and TiO2*,
SpringerBriefs in Molecular Science,
https://doi.org/10.1007/978-3-030-24032-5_4

of planetary orbits in the Solar System. At some point in the Earth's history, the Earth's temperature was low enough for water to condense and stay liquid to this day. Leaving now aside different scenarios which predict the condensation as early as a few tens of millions of years after the Earth formation or multiple condensations and evaporations, it is safe to declare that after the Late Heavy Bombardment, the Earth's water was condensed. Liquid water is an essential condition for the sustenance of life as we know it and even though theories exist of life originating in other solvents (Cassone et al. 2017), the origin of life in water seems as the most probable scenario. In turn, the presence of water implies that the Earth sustained oceans, lakes and rivers with a variety of organic and inorganic compounds dissolved in them. These oceans were also initially hot; the heating being driven by impact and volcanic activity (Plankensteiner et al. 2004a).

An important role was played by reactions of atmospheric gases reacting with available sources of energy, such as UV radiation, lightning or thermal energy in the form of volcanic heat. Atmospheric gases were supplied by impacts as well as outgassing from the surface. Such atmosphere therefore contained carbon oxides, nitrogen or sulphur oxides. The atmosphere maintained a greenhouse effect, which heated the planet up. The radiation incident into the atmosphere, in majority from the adjacent star, was inelastically scattered by the planetary surface and as a radiation with longer wavelengths is unable to pass through the atmosphere again. Instead, the photons were absorbed by greenhouse gases and re-emitted again to go for another round of scattering and absorption. All these processes also included a portion of non-radiative energy transfers where the trapped energy is released from the molecules, e.g. by collisions and so the radiation is gradually transformed into thermal energy.

When speaking of greenhouse gases nowadays, one usually has in mind gases such as CO, CO_2 and methane. From the physical and chemical point of view, all compounds that absorb infrared radiation to some extent can be considered greenhouse gases. Absorption in the infrared is the very quality that is measured in laboratories by IR spectroscopists. On a side note, this creates an observation barrier to space and telescopes are therefore often placed on the Earth's orbit or into higher elevations, such as the Atacama desert, which is not only on a plateau high above the sea level, but also very dry (water is a strong IR absorber).

On Earth, the atmospheric content of other infrared absorbing gases is very low they are therefore usually not considered. On exoplanets, however, the situation is different. If a hypothetical atmosphere is composed mainly of, for example, ethane, acetylene or HCN, it could maintain a greenhouse effect without the contribution of our classical greenhouse gases whatsoever.

Nevertheless, molecules that do not absorb infrared radiation are just as important for the planetary chemistry as the absorbing ones. On Earth, such role is played by oxygen. Most oxygen in the atmosphere was created by living organisms. From the first, however, small amount of oxygen was created by water photolysis or as a product of carbon dioxide photoreduction in water. This process is again enhanced by mineral surfaces, such as TiO_2 or possibly other semiconducting minerals. Oxygen created from these reactions, which are analogous to the laboratory reactions discussed throughout this book, may have been crucial for the overall oxidation state of the atmosphere and the evolution of life.

4.2   The Building Blocks of Life

Biological building blocks present on the early Earth must have been able to withstand strong UV radiation from the Sun without being destroyed. They must also have been stable in the hot prebiotic ocean and have the ability to carry information even is these conditions. To fulfil the basic role of replication of information, the molecules also had to replicate their structure/information in a defined and repeatable way. There exist only two groups, which are able to perform all the above-mentioned tasks—nucleic acids and proteins.

Some scientists think that nucleic acids, namely RNA would not be an ideal starting molecule in terms of stability in the primordial ocean compared to proteins. For example, Rode (1999) shows that peptides can carry and transfer information, replicate and be stable in high-salinity oceans, such as might have been present on the early Earth. This immediately suggests a need to find out about the synthesis of proteins' basic constituents—amino acids.

4.3   Synthesis of Amino Acids

The origin of experimental prebiotic chemistry dates to 1953, when Miller (1953) published an experiment showing that amino acids can be formed by lightning discharge in primitive reduced atmospheres containing gases such as CH_4, NH_3, CO and H_2. Miller detected several amino acids in his first paper. Later, his original samples were analysed again (Johnson et al. 2009) and many more amino acids were discovered (up to 33). In 2017, Ferus et al. (2017b) also showed that Miller must have synthesized nucleic acid bases by replicating Miller's experiments and using sensitive modern analysis techniques such as GC-MS. Miller's experiments inspired many researchers and in the past almost seventy years the experiments were varied in terms of the composition, energy inputs and experimental set-ups. Miller was later often criticized for the use of a reduced atmosphere because nowadays the long-term existence of such atmosphere on Earth seems improbable. It has been shown, however [e.g. Civiš et al. (2004)], that amino acids can be formed from neutral atmospheres as well. This neutral atmosphere is close to what is now expected as the secondary atmosphere on young Earth and consists mainly of CO_2, H_2O and N_2. This atmosphere could have been formed by volcanic outgassing, cometary and asteroid delivery and photochemical processes in synergy. Volcanic gases such as sulphur dioxide, nitrous oxide, nitric oxide or hydrogen sulphide could have been present, too. For this reason, Parker et al. propose that sulphur-containing amino acids can be produced in Miller-type experiments (Parker et al. 2011a, b). His experiments produced amino acids, but the atmosphere used (N_2, CH_4, CO) was criticized yet again for being too reducing to represent the atmosphere of the early Earth (Delano 2001). Sagan and Khare (1971) tested similar experiments representing the atmosphere of the Saturn's moon Titan. Such experiments (as repeated and varied many times after Imanaka et al. (2004)) produce so-called tholins. Tholins are polymers of C and N atoms with varied molecular weight that represent Titan's aerosol analogues of Earth.

Quite a number of experiments or simulations of early Earth or Titan's atmospheric chemistries produced prebiotic molecules. Most of them required liquid water during (Miller 1953) or after (Neish et al. 2010) production. In 2012, Hörst et al. (2012) showed that instead of water, the addition of oxygen in such atmospheres leads to the generation of amino acids or nucleic acid bases. Hence from CO_2 and UV radiation, it is possible to produce CO and CH_4 which, if it happens in a Titan-like atmosphere, may result in the production of amino acids despite the absence of water. Similar, the origin of biomolecules on Earth can be hypothesized in this way, namely in the upper atmosphere, where the UV radiation is not attenuated. Water may be crucial for the evolution of complex life, but it looks likely that the synthesis of their basic constituents will do without it.

It should be noted here that there exist many other possible sources of amino acids on the early Earth. Among the most prominent are, for example, impacting meteorites and extraterrestrial delivery in general. It has been shown that various meteorites, especially carbonaceous chondrites, contain amino acids and nucleic acid bases (Martins et al. 2008, 2015; Martins 2011). Other classes of organic substances have been reported as well (Kvenvolden et al. 1970; Pizzarello et al. 1991). Thomas et al. (2006) and (Jakschitz and Rode 2012) estimate that if every carbonaceous chondrite has 60 ppm of amino acids and 3% of all impacting meteorites were of this type, 6×10^{11} ton of amino acids could have been brought to Earth after it had cooled down to below water condensation point.

Aside from bringing in organic material, chondritic meteorites may play the role of catalysts in prebiotic reactions. Rotelli et al. (2016) claim that carbonaceous chondrites, which are the most primitive meteorites in the Solar System, are catalysed the productions of prebiotically relevant molecules from formamide in water. The products were all canonical nucleobases, amino acids and carboxylic acids. It is a question, however, whether formamide played the role of the parent molecules in the first place. This scenario is not limited to Earth, though, and if any planet in the universe possesses enough formamide. Prebiotic synthesis on carbonaceous chondrites may be possible there.

A different view is taken by scientists researching the chemistry of hydrothermal vents at the ocean floor. These vents are created in places where material or energy spouts from the underground, which is why the vents are often called 'black smokers'. The vents may serve as a source of amino acids as well (Huber and Wächtershäuser 1997). At high temperatures and pressures, life may have originated at the ocean bottom according to this view.

4.4 Synthesis of Peptides

Proteins are amino acid polymers created by condensation. One of the many possibilities to form proteins or peptides is condensation on the surface of clay minerals, such as montmorillonite, kaolinite, bentonite, silica or alumina. These minerals have a complex and often irregular structure and may serve as catalysts, shield the reaction products from decomposition or concentrate the products on their surface (Lahav et al. 1978; Cairns-Smith and Hartman 1986). The drawback of these reactions is the relatively low yield of proteins or peptides. Also, the reactions do not work very well for most of the proteinogenic amino acids. In spite of that, Bujdák and Rode (2001) and Flores and Bonner (1974) speculate that clays might have played a vital role in the early chemical evolution by adsorbing, elongating and protecting peptides.

4.5 Peptide Bond and TiO$_2$ Surface Catalysis

The condensation of amino acids is a basic step in the formation of peptides. In our body, the whole process is assisted by ribosomes, a template mRNA and a complex enzymatic machinery. But how did proteins form alone before any life truly begun? There are again many theories, some of them more likely than others.

One of the possible theories is condensation of amino acids on a TiO$_2$ surface. The mechanism of such condensation was investigated by Pantaleone et al. (2018), who performed PBE-D2* and PBE0-D2* periodic simulations on the TiO$_2$(101) anatase surface. Even though their work is prebiotically oriented, the authors rightly claim that this process can find its use in organic synthesis and in pharmaceutical synthesis in general.

In this reaction, the unsaturated Ti atoms adsorb water molecules and therefore act as a dehydration agent. Similar approach was taken by Martra et al. (2014), who succeeded in the synthesis of long glycine peptides by pumping glycine vapours over a TiO$_2$ surface in dry conditions. This experiment mimicked a small periodically drying pool. The experiments yielded poly-glycine chains up to 16 amino acids in length. Similar positive results were obtained with SiO$_2$. The experiment was repeated using hydroxyapatite instead of TiO$_2$ without success, indicating that the titanium dioxide and silicon dioxide play a partially selective role in the catalysis. Upon addition of water to the already formed poly-Gly chain, the formation of secondary structures was observed (both helices and β-sheets).

It should be noted that there are other scenarios which explain the formation of peptide bonds in prebiotic conditions and do not employ either TiO$_2$ or CO$_2$. As an example scenario, in 1989, Schwendinger and Rode (1989) theoretically investigated the salt-induced peptides formation of the peptide bond in NaCl solutions or divalent metal cations. They found out that because Na$^+$ above 3 M concentrations do not have a saturated first hydration shell, they can act as dehydrating agents as well and thus induce the formation of peptides. Most scenarios predict NaCl concentration in the primordial ocean about 0.5 M, i.e. roughly similar to what is in the oceans today. Reaching 3 M concentrations requires again the presence of a dehydrating pool or a lagoon with constant evaporation and replenishment by seawater, where the NaCl would be concentrated. Three years later, Schwendinger and Rode (1992) broadened their theory by the addition of Cu^{2+} ions. Divalent metal ions form complexes with the amino acids and lower the energy barrier of the reaction. Cu^{2+} ions were found to be most effective of all tested ions, because of the presence of Cl^- ions that bind to the Cu^{2+} centre as well. To give this theory a prebiotic relevance, rock formations of malachite and azurite from Precambrian times point towards the availability of Cu^{2+} ions (Nutman et al. 1996; Hofmann et al. 1999). The reaction itself was shown to work faster with α-amino acids compared to β-amino acids. Addition of glycine and diglycine also increases the yield (Plankensteiner et al. 2002). As Rode shows (Rode 1999), the condensation is not random and certain linkages are preferred. Similar sequences are preferred in archaebacterial membrane proteins, which make this reaction seemingly point towards just such an origin of first archaebacterial membranes.

4.6   Origin of Biohomochirality

This section does not directly concern CO_2 photocatalytic reduction or TiO_2 as a material, but is necessary for the thought transition between the peptide bond formation and parity violation, which will be discussed later.

The vast majority of amino acids in living systems are L-enantiomers and RNA/DNA uses D-sugars in the backbone. This quality is called homochirality and is a signature of living systems. The need for homochirality can be linked to the specific catalytic properties of enzymes that are selective to different enantiomers. Whether the enzymatic selection preceded the selectivity or vice versa or whether this is one of the chicken-egg questions is nowadays probably impossible to decide. The fact remains, however, that the enantiomeric purity of both amino acids and nucleic acids is critical for the functioning of an organism. In some cases, a single error in the sequence can lead to the collapse of a whole protein structure (Eberhard Krause et al. 2000). In the case of DNA, a single misread or a single error in the code can lead to a different amino acid incorporated into the protein structure, which again may lead to catastrophic events. Christopher J. Welch in 2001 published a paper (Welch 2001) on the origin of enantioenriched environments through conglomerate crystals. His results show that homochirality may have an abiotic origin and may not be the sole privilege of living systems. It is possible, however, that just such reactions gave rise to homochiral abiotic systems at a certain level of chemical complexity. These systems may have evolved to living systems which would inherently be homochiral. Another, less popular option is that a system that attained the ability to replicate was by pure statistical chance homochiral.

Yet another hypothesis for the origin of homochirality that is often seen puts the origin of homochirality beyond Earth. Dust in space scatters starlight and if the particles in the cloud are not spherical and are aligned in magnetic fields of star-forming regions, circularly polarized UV and IR light is emitted (Bailey et al. 1998). Circularly polarized light is known to enantioselectively affect reactions. For example, Flores et al. (1977) show that photolysis of racemic leucine produces up to 2.5% enantiomeric excess. The Murchison meteorite shows an excess of L-amino acids compared to D-amino acids. The effect of circularly polarized light can be, according to authors, the reason not only for the disparity in enantiomeric concentration in the Murchison meteorite, but also for the homochirality in living systems.

4.7   Parity Violation

Lee and Yang in 1956 predicted a parity conservation violation in the weak nuclear force (Lee and Yang 1956). Only a year later, the violation was experimentally proven. As a result, chiral molecules have different ground state energies and hence the reactivities. The energies depend on the weak nuclear interaction and the number of electrons in the atom/molecule in question. Electrons possess an inherent spin chirality and the heavier the atom the more electron in contains. Lighter atoms (H, N, C, O), therefore, have a very small difference in ground state energies. In heavier atoms, however, this difference is possible to measure (Jakschitz and Rode 2012) by optical rotatory dispersion. Addition of dimetal cations, such as Cu^{2+}, can lead to altered ground state energies of the forming amino acids (Jakschitz and Rode 2012). Different amino acid or peptide enantiomers could therefore exhibit different reactivities in the prebiotic pool of their origin and homochirality could in this way arise in living systems. Such hypothesis was tested by Plankensteiner et al. (2004b), who used pure L-Ala and D-Ala and monitored their behaviour in the salt-induced condensations. Firstly, the racemization was very slow. Secondly, L-alanine gave slightly higher yields of peptides than D-alanine. Different amino acids were tested and not all of them exhibited similar behaviour. Yet this is another effect that could be the reason for biohomochirality.

A different mechanism for the origin of biohomochirality was proposed by Fraser et al. (2011), who claim that vermiculite clay gels are sensitive to D-His and L-His in a different way and exhibit chiral selectivity in adsorption. Vermiculite is a clay mineral and contains Mg, Fe, Al, Si, O and H atoms. Different adsorption rates were observed by HPLC, namely after 10 h, 7.1% homochiral amplification was observed. The role of minerals may not be purely photocatalytical but selective adsorption on mineral surfaces could lead to biohomochirality as well.

The above-mentioned theories are only a selection of all published theories and were selected because they either include mineral surface catalysis, photochemistry of carbon dioxide or TiO_2 photocatalysis. What was the real reason for the origin of biohomochirality remains unknown. All the described models must have taken place in principle on young Earth as well as on any exoplanet imaginable. Some of them might have been interrupted by other reactions, may not be effective enough or simply happen in insignificant extents. One of the proposed models must have prevailed and be the reason for the observed L- and D-enantiomer disparity. Which it was, however, is nowadays impossible to decide.

Chapter 5
Applications of the CO_2 Photocatalytic Reduction

Abstract This chapter gives an overview of the state-of-the-art research progress in the topic of photocatalytic reduction of CO_2 in selected areas of application. The knowledge of the practical uses of the photocatalytic reduction process may be beneficial to give the book more context and to specify its place in the plethora of work that is available to any reader.

5.1 The Technical Aspect

There is currently a large amount of scientific and technological articles and books concerned with the study and practical uses of the photocatalytic reduction of CO_2 to methane and other hydrocarbons. Besides laboratory studies focused mainly on the chemistry and physics of the catalyst and the process itself, there are also publications aimed at the construction of model or functional devices. These could serve in future as first prototypes for mass-produced apparatuses for the solar energy capture and carbon dioxide reduction in places with sufficient solar irradiation.

This book is focused in a slightly different way. The main aim is to point at several exceptional features of semiconductor photocatalysts and their behaviour in the presence of CO_2, which are largely or completely neglected in the presented literature. For this reason, this chapter will be only a review of the most important or most intensely studied applications and should give the reader a brief illustration of the possible applications of the photocatalytic reduction of CO_2 in various areas of technological research. Before we immerse ourselves into the review, it is necessary to show some basic facts and principles in the semiconductor chemistry and in the photocatalytic reduction of CO_2.

© The Author(s), under exclusive license to Springer Nature Switzerland AG 2019
S. Civiš et al., *The Chemistry of CO_2 and TiO_2*,
SpringerBriefs in Molecular Science,
https://doi.org/10.1007/978-3-030-24032-5_5

5.2   Basic Principles of the Photocatalytic Reduction of CO_2 on Semiconductors

Photocatalytic reduction of CO_2 to a certain extent mimics photosynthesis, which is a natural process in photosynthetic organisms. In this process, CO_2 is captured and through chemical conversion using solar energy is converted to molecules with carbon in a more reduced oxidation state. In plants, this is typically glucose or a different sugar. In the photocatalytic reduction of CO_2 in laboratory conditions, this may be methane, methanol, formaldehyde, formic acid or other (Noël 2017), as shown in Fig. 5.1. These products can be used for different purposes, such as chemical synthesis, as a fuel or in the food industry as reagents.

Photosynthesis is a combination of water oxidation and carbon dioxide reduction. The process is a multi-reaction process and involves both light-powered (light) and light-independent (dark) reactions (Lingampalli et al. 2017).

The photocatalytic reduction, schematically shown in Fig. 5.2, consists of three steps (Rao and Lingampalli 2016):

1. Absorption of light by the semiconductor; $(hv > E_{bg})$, where E_{bg} is the width of the band gap.
2. Separation and migration of charge carriers.
3. Reactions at the surface.

These three processes can be separately studied and enhanced.

In step 1, the photocatalyst is illuminated and the light is absorbed in the material. The condition for the absorption of light in the semiconductor is that the energy of the photon, hv, must be equal to or higher than the energy of the band gap, i.e. the photon

Fig. 5.1 Photosynthesis is a process in which photosynthetic organisms reduce CO_2 and oxidize water

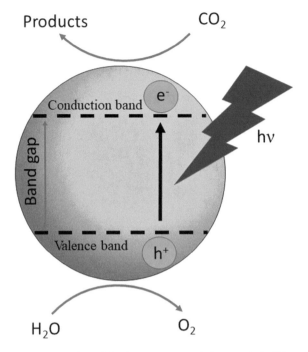

Fig. 5.2 Schematic representation of the three processes of photocatalytic CO_2 reduction

must be able to excite an electron from the valence band to the conduction band. The reduction of CO_2 is a process which requires energy and in terms of potential is an uphill reaction. Water splitting is in terms of potential a downhill reaction. These reactions should therefore be coupled to keep the redox balance in the system. As is described by Lingampalli et al. (2017), the conduction band edge is negative relative to the redox potential of CO_2 and facilitates the transfer of electrons towards CO_2. On the other side, the valence band edge is positive relative to water redox potential and allows for the transfer of holes to the water molecule. Both these processes, or the coupling, are thermodynamically favoured as opposed to direct CO_2 reduction. Simply put, electrons from the conduction band reduce CO_2 and holes in the valence band oxidize water.

The photocatalytic reduction is a network of reactions, some of which were already discovered (Sharma et al. 2008; Costentin et al. 2013; de Richter et al. 2013; Li et al. 2019). For these reactions, it is possible to determine the redox potentials. The potentials were determined against the standard hydrogen electrode at pH 7 (White et al. 2015).

$$CO_2 + e^- \rightarrow CO_2^{\cdot-} \quad -1.850 \text{ V} \tag{5.1}$$

$$CO_2\ (g) + H_2O\ (l) + 2e^- \rightarrow HCOO^-(aq) + OH^-(aq) \qquad -0.665\ V \qquad (5.2)$$

$$CO_2\ (g) + H_2O\ (l) + 2e^- \rightarrow CO(g) + 2OH^-(aq) \qquad -0.521\ V \qquad (5.3)$$

$$CO_2\ (g) + 3H_2O\ (l) + 4e^- \rightarrow HCHO(l) + 4OH^-(aq) \qquad -0.485\ V \qquad (5.4)$$

$$CO_2\ (g) + 5H_2O\ (l) + 6e^- \rightarrow CH_3OH + 6OH^-(aq) \qquad -0.399\ V \qquad (5.5)$$

$$CO_2\ (g) + 6H_2O + 8e^- \rightarrow CH_4 + 8OH^-(aq) \qquad -0.246\ V \qquad (5.6)$$

$$2H_2O\ (l) + 2e^- \rightarrow H_2(g) + 2OH^-(aq) \qquad -0.414\ V \qquad (5.7)$$

$$2H_2O\ (l) \rightarrow O_2\ (g) + 4H^+(aq) + 4e^- \qquad +0.816\ V \qquad (5.8)$$

The one-electron reaction (5.1) has a large negative reduction potential relative to the conduction band. This was explained by the need for a change in hybridization from sp^2 to sp^3. On the other hand, multielectron reactions/reaction chains, such as the formation of HCHO, CH_3OH or HCOOH are small. These reactions are therefore much more feasible. Since the water-to-oxygen reaction has a redox potential 0.816 V, coupling with reactions with lesser potential is preferable (White et al. 2015).

From this, the free energy change, ΔG, values can be obtained for each reaction from $\Delta G = -nFE$, where n is the molar amount of substance, F is the Faraday constant and E is energy. This implies that the ΔG in the reactions 5.1–5.8 is positive. $\Delta G > 0$ means that the reaction is not spontaneous. The smallest ΔG can be obtained for HCOOH formation, whereas the highest for CH_4 formation. The full order is, from smallest to highest, HCOOH < CO < HCHO < CH_3OH < CH_4. The splitting of water is spontaneous in terms of ΔG. It does not mean, however, that the reaction would be spontaneous in terms of kinetics, since there exists a significant activation barrier. The full mechanism, the sequence of reactions and the fate of the electron-hole pair are still widely discussed (White et al. 2015).

Similar to the oxygen atom exchange described in the previous chapters, the efficiency of photocatalysts is affected by the mineral morphology, facets, particle size, surface defects, lattice structure and the chemical environment. The activity was also shown to be affected by pH, solvent, surface OH^- and H_3O^+ groups and dopants. A change in pH changes the reaction rate of CO_2 and OH^-, thus disturbing the balance between these species (Gattrell et al. 2006) and modifying the reaction rates of the other reactions. Different chemical species are adsorbed to a different extent, are available to a different extent and possible products or intermediates may react through side reactions. The reaction is also affected, mainly in the gas phase, by pressure, temperature and water content (Endrődi et al. 2017).

5.3 Examples Based on Doped TiO$_2$

One possibility of increasing the effectivity of the photocatalytic reduction of CO_2 over semiconductor photocatalysts is doping of the material. In general, the dopant adds energy levels to the band gap of the material and in this way facilitates the electron excitation and photon capture. Tahir et al. (2016) explored investigated the direct photocatalytic reduction of CO_2 using water or H_2. The authors used NiO/In_2O_3 doped TiO$_2$ catalysts. Specific pore diameter and density were observed when NiO and In_2O_3 were loaded onto TiO$_2$ during the crystal growth phase. The In and Ni atoms, now incorporated into the TiO$_2$ structure, promoted charge separation (electron-hole pair) in the catalyst and at the same time suppressed the electron-hole pair recombination. Methane was produced in this experiment when H_2O was used as a reducing agent. The authors determined that 1 wt% of NiO and 3 wt% if In_2O_3 was the optimal addition to the TiO$_2$ crystal to obtain maximum methanogenesis efficiency (240 μmol l^{-1} h^{-1}). This was 6.5 times more efficient than pure TiO$_2$. The reaction was also attempted with H_2 as the reductant 208 μmol l^{-1} h^{-1} of CH$_4$ was produced. At the same time, the production rate of CO was 243 μmol l^{-1} h^{-1}. Therefore, water as the reducing agent is better for CH$_4$ production. However, H_2 is more effective for the use up of CO_2.

Different doping in a similar system of CO_2, H_2O and TiO$_2$ was studied by Amin, Tahir and Tahir (Tahir et al. 2018). The main products of this experiment were CO, H_2 and C_2H_6. At 5 wt% doping by La, the amount of generated CO from a CO_2–CH$_4$ system was 9.6 times higher than on pure TiO$_2$. Addition of water to this system increased the CO yield 37 fold. The increase was again attributed to the La^{3+} facilitating charge separation and preventing recombination of the electron-hole pair.

The authors also presented a PTOP value (photocatalytic turnover productivity), which puts into relation the photon energy consumption and the catalyst effectivity. In the case of CO, the PTOP in the CO_2–CH$_4$ system was 3.83 fold lesser than in the CO_2–H_2O–CH$_4$ system. H_2 and C_2H_6, however, had the PTOP 1.2× and 2.1× higher in the CO_2–CH$_4$ system compared to the CO_2–H_2O–CH$_4$ system. La-doped TiO$_2$ also had larger stability in the long term when the CO_2–H_2O–CH$_4$ system was used. H_2O therefore enhances CO production and increases the stability and photochemistry of the system in the long run.

Nitrogen-doped TiO$_2$ nanotube arrays were also tested as photocatalysts in the methanogenesis (Delavari et al. 2016). The arrays were manufactured by the anodization method and the effects of UV irradiation power, starting mixture composition (CO_2, CH$_4$, N_2) and the distance from the UV lamp were evaluated by the response surface methodology. The nanotubes were 3–50 nm in diameter and ordered. In the CO_2, CH$_4$ and N_2 system, the optimal conditions (250 W UV light source, 10% of CO_2 in the mixture and 2 cm from the UV lamp) yielded H_2 and CO. The authors used a Langmuir-Hinshelwood kinetic model and incorporated photocatalytic adsorptive reduction and oxidation and observed that the model fitted well to the observed data.

Both CO$_2$ and CH$_4$ were competitively activated and their ratio in the initial mixture has an effect on the product mixture ratio.

5.3.1 An Overview of Important Reviews and Books Concerned with CO$_2$ Photocatalytic Reduction

Further information on the topic of industrial applications of the photocatalytic reduction of CO$_2$ can be found in the following publications.

A review by Wang et al. (2017) summarizes other techniques and photocatalysts present in the literature and splits them into categories according to what energy range they capture from the solar radiation and what applications they can be used for.

Huang et al. (2018) in their review present used optofluid microreactors for water splitting coupled with CO$_2$ photocatalytic reduction. This is a novel approach which takes inspiration from the natural photosynthesis, which is itself an optofluid process. The reduction of the reactor size to microscale is an interesting feat which may help implement the photocatalytic reduction in places and sites where previously it would seem impossible. In the review, a brief introduction of the artificial photosynthesis is given and then recent progress in water splitting, CO$_2$ reduction and coenzyme regeneration is presented.

Nguyen and Wu (2018) summarize the recent developments in the design and construction of photoreactors. Photoreactors are devices for solar energy conversion in general and the development of commercially available photoreactors is the ultimate goal of the research in the field. Currently, there exist several types of photoreactors for the photocatalytic reduction of CO$_2$. The most important and promising are optical-fibre photoreactors, monolith photoreactors and internally illuminated monolith reactors. Designing reactors with mixed use (CO$_2$ photocatalytic reduction coupled with water splitting) is expected to be the major branch of research in the near future. These reactors could utilize CO$_2$ and solar energy and at the same time use up wastewater.

A review by Ola and Maroto-Valer (2015) presents contemporary material design and reactor engineering in the photocatalytic reduction of CO$_2$. The review described general aspects of the problematic, but the main emphasis is put on the material aspects and engineering aspects. TiO$_2$-based catalysts are described, with special attention paid to improvement of selectivity and conversion rates.

Enrődi et al. (2017) present and review the technique of continuous-flow electroreduction of CO$_2$. Enrődi et al. write that electrochemical reduction in electrochemical reactors might be the only way how to scale up the reaction for industrial applications and therefore this method should be paid more attention to. The article then discusses the electrolysers and their parameters. Microfluidic and membrane-based electrochemical cell and their operational conditions are discussed as well. Based on this description, benchmarking methodology for the evaluation of CO_2 electrolysers is given and in the end, photoelectrochemical reduction is discussed.

Protti et al. (2014) in their review present an interesting view on the photocatalytic reduction of CO_2. According to the authors, 12–13% of the current information value on the photocatalytic reduction of CO_2 is presented in patents. Their review therefore focuses on patent literature on the topic. The authors describe how a redshift in the absorption of materials has been achieved, but did not always lead to better photocatalytic properties. The redshift is important to increase the utilization of solar radiation, but the energy of the individual photons is lower and the chemical energy they are able to release is therefore smaller. Technical details of the charge recombination, catalyst corrosion and water splitting are discussed as well as the hydrogen production through sacrificial hydrogen donors.

ChemSusChem presented a series of articles where its board members present and discuss the most important articles concerned with sustainability. In one of these, North (2019) writes about the concept of carbon dioxide biorefinery and reviews contemporary carbon dioxide utilization techniques for the production of fuels and chemicals.

Colmenares and Xu in 2016 published a book on the photocatalytic reduction of CO_2 called 'Heterogenous photocatalysis: From fundamentals to green applications' (Colmenares and Xu 2016). The book presents basic principles of photocatalysis and then turns its attention to green applications using both inorganic and organic catalysts. TiO$_2$-based composites are also discussed and the latest developments (up to 2016) are presented. From TiO$_2$, the authors move to a description of possible alternatives in photocatalysis. In a more general context of photocatalysis (not limited to CO_2 photoreduction), heterogenized polyoxometalate materials and photoreactor design are presented as well.

Lichtfouse, Schwarzbauer and Robert in their book 'Hydrogen production and remediation of carbon and pollutants' give detailed description of hydrogen production from biomass (Lichtfouse et al. 2015). Three chapters of this book are devoted to the CO_2 atmospheric production and content issue and possible solutions are described. CO_2 capture and sequestration in terrestrial biomass are described to give more context to the topic.

'Nanomaterials for photocatalytic chemistry' by Yugang Sun (Yan et al. 2017) focuses on the use of nanomaterials and solar energy for the catalysis of chemical reactions such as water splitting, CO_2 reduction and selective alcohol oxidation and epoxidation. Nanomaterial design and synthesis are covered within this book. Perspective for near-future advances is also given, and outlooks in chemical conversion using nanomaterials and photocatalysis are shown.

Timothy Noël in a book 'Photochemical processes in continuous-flow reactors' presents an overview of microreactors and their actual technological and chemical uses (Noël 2017). Case studies for specific practical uses are described and on these, the advantages and drawbacks of various approaches are shown and discussed. The case studies include, for example, continuous-flow photoreactor design, modelling of photochemical transformation, photon transfer phenomena, scaling up strategies, the use of continuous-flow photocatalysis, industrial applications and more.

Fujishima, Hashimoto and Watanabe in their 2001 book 'TiO$_2$ photocatalysis: fundamentals and applications' describe many commercially available technologies related to TiO$_2$ and its use as a solar light-powered photocatalyst (Fujishima et al. 1999). The main topics covered are cleaning technology, anti-bacterial effects, self-cleaning and self-sterilizing materials, air purification, water purification, cancer therapy and photocatalysis resembling photosynthesis. All these applications were to a certain extent and with certain success tested and are presented in this book.

Chapter 6
Conclusion

Abstract In this last chapter, we summarize the findings and key points presented in this book for the convenience of the reader. This chapter is divided into sections, each describing a separate problematic discussed in this book. The topics go in the order in which they were presented in this book.

6.1 Oxygen Isotope Exchange Between Carbon Dioxide and TiO_2

The totally ^{18}O—isotope exchanged titanium dioxide, $Ti^{18}O_2$ (anatase) was prepared from $TiCl_4$ and $H_2^{18}O$.

The $Ti^{18}O_2$ anatase sample calcined at 450 °C shows a very high rate of spontaneous exchange of oxygen isotopes with $C^{16,16}O_2$. The surface layer of the vacuum-annealed mineral is strewn with defects, which are a mixture of Ti^{4+} and Ti^{3+} centres. In the process of the exchange, $C^{16}O_2$ molecule is adsorbed on the surface, where, as calculations show, it forms a bidentate CO_3^{2-} ion. The three oxygens in the carbonate are symmetrically equivalent and the release of $C^{18,16}O_2$ is probable (2/3 probability of its release vs. 1/3 probability of the release of the unmodified $C^{16}O_2$). Repeating this step, $C^{18}O_2$ is in the end formed. Mechanism of oxygen exchange between molecule and surface has been analysed by DFT calculations with inclusion of corrections for long-range dispersion, with and without on-site Coulomb interactions. The CO_2 adsorption site has been identified at oxygen defect sites, Vo, where the molecule can be adsorbed in four different configurations L(1), L(2), B(1) and B(2). The intermediate configuration is described as C(0) and is a carbonate bidentate ion. The ion is at the same time bonded to Ti(5f) and O(2f) surface atoms and it allows for direct release of the CO_2 molecule with an exchanged oxygen isotope.

If calcined at 200 °C, after the irradiation with UV laser, additional rotation-vibration bands of methane, acetylene and water were identified in the gaseous phase.

The water molecules which are released or created in the gas phase show that no exchange of oxygen atoms ^{18}O from the solid phase $Ti^{18}O_2$ take place.

© The Author(s), under exclusive license to Springer Nature Switzerland AG 2019
S. Civiš et al., *The Chemistry of CO2 and TiO2*,
SpringerBriefs in Molecular Science,
https://doi.org/10.1007/978-3-030-24032-5_6

A small amount of $C^{16}O$ generated from the gaseous $C^{16}O_2$ by laser irradiation remains unchanged in the gas phase. The isotope exchange does not take place between the oxygen structure, because CO bonds directly to the Ti atoms.

6.2 Oxygen Isotope Exchange Between Carbon Dioxide and Synthetic and Natural Minerals

It is therefore obvious that the rate of oxygen exchange is very sensitive to the sample pretreatment. Systematic examination of CO_2 interaction with clay, natural anatase TiO_2, natural rutile, clays from the Sokolov Coal Basin, and several commercial and natural materials showed that ^{16}O and ^{18}O are commonly exchanged between the mineral surface and the gas phase CO_2. Such O-exchange process reaches very high values typical for nanoparticulated synthetic material.

The oxygen exchange activity was tested for other minerals as well. First, the effect of crystal structure was explored by performing the photocatalytic reduction on rutile. Having found out that rutile is active, the effect of annealing was tested as well. With the increasing temperature, the number of oxygen defect sites increases and the oxygen exchange is faster.

This is not true in the case of clays and montmorillonite. Natural clays and synthetic montmorillonite were tested and even though both exhibited oxygen exchange activity, when annealed at 450 °C, the ability to interact with the $C^{18,18}O_2$ was reduced. This was explained by the effect of sintering, which increases particle size and reduced the effective surface area of the sample. This finding was verified by CO_2 adsorption capacities measurement.

Last, the oxygen mobility between CO_2 and basalt, calcite and siderite was tested. All three materials exhibited some activity, but its rate did not depend either on the surface area or the annealing temperature (oxygen defect surface density), or rather, no trend or correlation could be found. From this, we inferred that the mobility of oxygen and exchange activity depends on the chemical environment as well.

So there are in total, three main effects that influence the oxygen mobility between CO_2 and the solid phase. These are the oxygen defect surface density, surface area of the mineral and chemical properties of the material in general. Only synergy of these effects will maximize the efficiency of the minerals in the CO_2 interaction activity.

6.3 Photocatalytic Transformation of CO_2 to CH_4

Results on the photocatalytic reduction of CO_2 to CO and methane using UV radiation show that on planets of terrestrial type a periodic partial and temporary reduction of the atmosphere is possible. This process must be considered in planetary models and models of early atmosphere composition on Earth and other planets. Special accent is exerted towards Mars, which is our planetary neighbour and has many features similar to Earth, and the Saturn's moon Titan, which shows high concentrations of methane on its surface. The conversion of such atmosphere has been tested using a 350 nm UV lamp and mineral catalysts such as anatase, montmorillonite or the Nakhla meteorite.

Apart from methane, the experiment produced perchlorates, chlorates and methylchlorides as products created due to the presence of HCl. These were detected by either XPS or GC-MS. This leads us to the conclusion that these molecules can be continually produced on Mars and their origin can be explained by this continual production and accumulation in the soil. This mechanism should be powerful enough to explain the 0.5% perchlorate content in the Martian soil. It also predicts that this species can be found from 5 to 50 cm depth beneath the surface.

Also, the mechanism predicts a trend in the production of CO, which should behave in a similar manner to methane and fluctuate seasonally. The relative stability of CO and possible reverse reactions of CH_4 oxidation will smear the fluctuations significantly, but still, we believe that their presence could be detected.

A future experiment on a next mission to Mars should systematically search for these fluctuations and compare them to the local mineral composition and weather.

Moreover, creation of such a reduced atmosphere, at least locally and temporarily, may serve as a first step towards creation of biologically active molecules. This atmosphere can be transformed by exposure to high energy density environment, such as laser plasma or asteroid impact, and yield vast quantities of HCN, NH_3, C_2H_2 or ethene. The mechanism is dominated by radical chemistry of $\cdot$CN, $\cdot$NH, water and vibrationally excited CO molecules. Using gas chromatography-mass spectrometry, glycine and RNA nucleic acid bases have been detected after the transformation.

6.3.1 The CO_2 Cycle

Carbon dioxide-rich atmosphere can be converted to an atmosphere containing CH_4 and CO. This atmosphere has better potential for prebiotic synthesis. Also, some atmospheric features can be predicted from the mechanism, such as CO and CH_4 fluctuations. These are all testable predictions. It should be noted that the conversion may be partial and temporary and yet sufficient to drive the described processes. The experiments also show the creation of perchlorates, which were found on Mars and a mechanism of their creation and accumulation is proposed here.

The reduced or partially reduced atmosphere serves as a starting atmosphere from which prebiotic molecules are synthesized. These include amino acids, nucleic acid bases and sugars. Due to the occurrence of a cataclysmic event, such as asteroid impact, frequent lightning or massive volcano eruption, these molecules would be burned back to CO_2. In this way, the CO_2 cycle is complete.

References

H. Abdullah, M.M.R. Khan, H.R. Ong, Z. Yaakob, Modified TiO_2 photocatalyst for CO_2 photocatalytic reduction: an overview. J. CO_2 Utilization **22**, 15–32 (2017). https://doi.org/10.1016/j.jcou.2017.08.004

M. Aresta, A. Dibenedetto, Utilisation of CO_2 as a chemical feedstock: opportunities and challenges. Dalt. Trans. 2975–2992 (2007). https://doi.org/10.1039/b700658f

R. Atkinson, D.L. Baulch, R.A. Cox et al., Evaluated kinetic and photochemical data for atmospheric chemistry supplement—IV—Iupac subcommittee on gas kinetic data evaluation for atmospheric chemistry. J. Phys. Chem. Ref. Data **21**, 1125–1568 (1992). https://doi.org/10.1063/1.555918

S.K. Atreya, P.R. Mahaffy, A.-S. Wong, Methane and related trace species on Mars: origin, loss, implications for life, and habitability. Planet. Space Sci. **55**, 358–369 (2007). https://doi.org/10.1016/j.pss.2006.02.005

J. Bailey, A. Chrysostomou, J.H. Hough et al., Circular polarization in star-formation regions: implications for biomolecular homochirality. Science **281**, 672–674 (1998)

A. Bar-Nun, V. Dimitrov, Methane on Mars: a product of H_2O photolysis in the presence of CO. Icarus **181**, 320–322 (2006). https://doi.org/10.1016/J.ICARUS.2005.11.023

J.K. Beatty, C.C. Petersen, A. Chaikin (eds.), *The New Solar System*, 4th edn. (Cambridge University Press, Cambridge, UK, 2015)

P. Beckmann, A. Spizzichino, *The scattering of electromagnetic waves from rough surfaces* (Artech House, Norwood, MA, 1987), p. c1963, ISBN: 0890062382

K. Biemann, J. Oro, P. Toulmin et al., The search for organic substances and inorganic volatile compounds in the surface of Mars. J. Geophys. Res. **82**, 4641–4658 (1977)

P. Bogdanoff, N. Alonso-Vante, A kinetic approach of competitive photoelectrooxidation of HCOOH and H_2O on TiO_2 anatase thin-layers via online mass detection. J Electroanal. Chem. **379**, 415–421 (1994). https://doi.org/10.1016/0022-0728(94)87165-5

P. Boháček, J. Prachařová, S. Civiš et al., Composition of phases and phase mixtures of the Bi-Sr-Ca-Cu-O system. Phys. C **171**, 108–120 (1990). https://doi.org/10.1016/0921-4534(90)90462-N

G. Bravo-Perez, J.R. Alvarez-Idaboy, A.G. Jimenez, A. Cruz-Torres, Quantum chemical and conventional TST calculations of rate constants for the OH plus alkane reaction. Chem. Phys. **310**, 213–223 (2005). https://doi.org/10.1016/j.chemphys.2004.10.031

C.A.M. Brenninkmeijer, C. Janssen, J. Kaiser et al., Isotope effects in the chemistry of atmospheric trace compounds. Chem. Rev. **103**, 5125–5161 (2003). https://doi.org/10.1021/cr020644k

D. Brown, Phoenix Mars team opens window on scientific process. https://www.nasa.gov/mission_pages/phoenix/news/phoenix-20080805.html. Accessed 26 Apr 2018

J. Bujdák, B.M. Rode, Activated alumina as an energy source for peptide bond formation: consequences for mineral-mediated prebiotic processes. Amino Acids **21**, 281–291 (2001). https://doi.org/10.1007/s007260170014

A.G. Cairns-Smith, H. Hartman (eds.), *Clay Minerals and The Origin of Life* (Cambridge University Press, Cambridge, UK, 1986)

Carbon Dioxide Information Analysis Center, Environmental Sciences Division, Oak Ridge National Laboratory, Tennessee US (2019) CO_2 emissions (kt). https://data.worldbank.org/indicator/EN.ATM.CO2E.KT. Accessed 11 Apr 2019

B.L. Carrier, S.P. Kounaves, The origins of perchlorate in the Martian soil. Geophys. Res. Lett. **42**, 3739–3745 (2015). https://doi.org/10.1002/2015GL064290

G. Cassone, J. Sponer, F. Saija et al., Stability of 2′,3′ and 3′,5′ cyclic nucleotides in formamide and in water: a theoretical insight into the factors controlling the accumulation of nucleic acid building blocks in a prebiotic pool. Phys. Chem. Chem. Phys. **19**, 1817–1825 (2017). https://doi.org/10.1039/c6cp07993h

D.C. Catling, M.W. Claire, K.J. Zahnle et al., Atmospheric origins of perchlorate on Mars and in the Atacama. J. Geophys. Res. Planets **115**, E00E11 (2010). https://doi.org/10.1029/2009je003425

G. Centi, S. Perathoner, Towards solar fuels from water and CO_2. ChemSusChem **3**, 195–208 (2010). https://doi.org/10.1002/cssc.200900289

X. Chang, T. Wang, J. Gong, CO_2 photo-reduction: insights into CO_2 activation and reaction on surfaces of photocatalysts. Energy Environ. Sci. **9**, 2177–2196 (2016). https://doi.org/10.1039/C6EE00383D

H.Z. Cheng, A. Selloni, Energetics and diffusion of intrinsic surface and subsurface defects on anatase TiO(2)(101). J. Chem. Phys. **131**, 54701–54703 (2009). https://doi.org/10.1063/1.3194301

D.M. Christodoulou, D. Kazanas, Conundrums and constraints concerning the formation of our solar system—an alternative view. 2017–2020 (2019)

C. Chyba, C. Sagan, Endogenous production, exogenous delivery and impact-shock synthesis of organic molecules—an inventory for the origin of life. Nature **355**, 125–132 (1992). https://doi.org/10.1038/355125a0

S. Civiš, A. Walters, M.Y. Tretyakov et al., Submillimeter-wave spectral lines of negative ions (SH- and SD-) identified by their Doppler shift. J. Chem. Phys. **108**, 8369–8373 (1998)

S. Civiš, L. Juha, D. Babankova et al., Amino acid formation induced by high-power laser in CO2/CO-N-2-H2O gas mixtures. Chem. Phys. Lett **386**, 169–173 (2004). https://doi.org/10.1016/j.cplett.2004.01.034

S. Civiš, M. Ferus, P. Kubát et al., Oxygen-isotope exchange between CO_2 and Solid (TiO_2)-O-18. J. Phys. Chem. C **115**, 11156–11162 (2011). https://doi.org/10.1021/jp201935e

S. Civiš, M. Ferus, M. Zukalova et al., Photochemistry and gas-phase FTIR spectroscopy of formic acid interaction with anatase (TiO_2)-O-18 nanoparticles. J. Phys. Chem. C **116**, 11200–11205 (2012). https://doi.org/10.1021/jp303011a

S. Civiš, M. Ferus, M. Zukalová et al., The application of high-resolution IR spectroscopy and isotope labeling for detailed investigation of TiO_2/gas interface reactions. Opt. Mater. (Amst) **36**, 159–162 (2013). https://doi.org/10.1016/j.optmat.2013.04.009

S. Civiš, M. Ferus, J.E.J. Sponer et al., Room temperature spontaneous conversion of OCS to CO_2 on the anatase TiO_2 surface. Chem. Commun. **50**, 7712–7715 (2014). https://doi.org/10.1039/c4cc01992j

S. Civiš, M. Ferus, M. Zukalová et al., Oxygen atom exchange between gaseous CO_2 and TiO_2 Nanoclusters. J. Phys. Chem. C **119**, 3605–3612 (2015). https://doi.org/10.1021/jp512059b

S. Civiš, M. Ferus, A. Knížek et al., Photocatalytic transformation of CO_2 to CH_4 and CO on acidic surface of TiO_2 anatase. Opt. Mater. (Amst) **56**, 80–83 (2016). https://doi.org/10.1016/j.optmat.2015.11.015

S. Civiš, A. Knížek, P. Kubelík et al., Spontaneous oxygen isotope exchange between carbon dioxide and oxygen containing minerals (Do the minerals "breathe" CO_2?). J. Phys. Chem. C **120**, 508–516 (2016). https://doi.org/10.1021/acs.jpcc.5b11306

S. Civiš, A. Knížek, O. Ivanek et al., Origin of methane and biomolecules from a CO_2 cycle on terrestrial planets. Nat. Astron. **1**, 721–726 (2017). https://doi.org/10.1038/s41550-017-0260-8

S. Civiš, A. Knížek, P.B. Rimmer et al., Formation of methane and (Per)chlorates on Mars. ACS Earth Space Chem. **3**, 221–232 (2019). https://doi.org/10.1021/acsearthspacechem.8b00104

CO_2.earth CO_2—Earth. https://www.co2.earth/. Accessed 26 Nov 2018

C.S. Cockell, Biological effects of high ultraviole radiation on early Earth—a theoretical evaluation. J. Theor. Biol. **193**, 717–729 (1998). https://doi.org/10.1006/jtbi.1998.0738

J.C. Colmenares, Y.-J. Xu (eds.), *Heterogeneous Photocatalysis* (Springer, Berlin Heidelberg, Berlin, Heidelberg, 2016)

J.C. Corchado, J. Espinosa-Garcia, O. Roberto-Neto et al., Dual-level direct dynamics calculations of the reaction rates for a Jahn-Teller reaction: Hydrogen abstraction from CH(4) or CD(4) by O((3)P). J. Phys. Chem. A **102**, 4899–4910 (1998). https://doi.org/10.1021/jp980936i

C. Costentin, M. Robert, J.-M. Savéant, Catalysis of the electrochemical reduction of carbon dioxide. Chem. Soc. Rev. **42**, 2423–2436 (2013). https://doi.org/10.1039/C2CS35360A

A. Davila, D. Willson, J.D. Coates, C.P. Mckay, Perchlorate on Mars: a chemical hazard and a resource for humans. Int. J. Astrobiol. **12**, 321–325 (2013). https://doi.org/10.1017/S1473550413000189

R.K. de_Richter, T. Ming, S. Caillol, Fighting global warming by photocatalytic reduction of CO_2 using giant photocatalytic reactors. Renew. Sustain. Energy Rev. **19**, 82–106 (2013). https://doi.org/10.1016/j.rser.2012.10.026

J.W. Delano, Redox history of the earth's interior since ~ 3900 Ma: implications for prebiotic molecules. Orig. Life Evol. Biosph. **31**, 311–341 (2001). https://doi.org/10.1023/A:1011895600380

S. Delavari, N.A.S. Amin, M. Ghaedi, Photocatalytic conversion and kinetic study of CO2 and CH4 over nitrogen-doped titania nanotube arrays. J. Clean Prod. **111**, 143–154 (2016). https://doi.org/10.1016/J.JCLEPRO.2015.07.077

A. Demont, S. Abanades, Solar thermochemical conversion of CO_2 into fuel via two-step redox cycling of non-stoichiometric Mn-containing perovskite oxides. J. Mater. Chem. A **3**, 3536–3546 (2015). https://doi.org/10.1039/c4ta06655c

A. Dhakshinamoorthy, S. Navalon, A. Corma, H. Garcia, Photocatalytic CO_2 reduction by TiO_2 and related titanium containing solids. Energy Environ. Sci. **5**, 9217 (2012). https://doi.org/10.1039/c2ee21948d

C. Di Valentin, G. Pacchioni, A. Selloni, Reduced and n-type doped TiO_2: nature of Ti^{3+} species. J. Phys. Chem. C **113**, 20543–20552 (2009). https://doi.org/10.1021/jp9061797

N.M. Dimitrijevic, B.K. Vijayan, O.G. Poluektov et al., Role of water and carbonates in photocatalytic transformation of CO_2 to CH_4 on titania. J. Am. Chem. Soc. **133**, 3964–3971 (2011). https://doi.org/10.1021/ja108791u

N.M. Dimitrijevic, I.A. Shkrob, D.J. Gosztola, T. Rajh, Dynamics of interfacial charge transfer to formic acid, formaldehyde, and methanol on the surface of TiO_2 nanoparticles and its role in methane production. J. Phys. Chem. C **116**, 878–885 (2012). https://doi.org/10.1021/jp2090473

B.L. Ehlmann, C.C. Edwards, Mineralogy of the Martian Surface, in *Annual Review of Earth and Planetary Sciences*, ed. by R. Jeanloz (2014), pp. 291–315

T. Encrenaz, T. Fouchet, R. Melchiorri et al., Seasonal variations of the martian CO over Hellas as observed by OMEGA/Mars express. Astron. Astrophys. **459**, 265–270 (2006). https://doi.org/10.1051/0004-6361:20065586

B. Endrődi, G. Bencsik, F. Darvas et al., Continuous-flow electroreduction of carbon dioxide. Prog. Energy Combust. Sci. **62**, 133–154 (2017). https://doi.org/10.1016/J.PECS.2017.05.005

J. Felipe Montoya, J. Peral, P. Salvador, Surface chemistry and interfacial charge-transfer mechanisms in photoinduced oxygen exchange at O-2-TiO_2 interfaces. ChemPhysChem **12**, 901–907 (2011). https://doi.org/10.1002/cphc.201000611

J.P. Ferris, A.R.J. Hill, R. Liu, L.E. Orgel, Synthesis of long prebiotic oligomers on mineral surfaces. Nature **381**, 59–61 (1996). https://doi.org/10.1038/381059a0

M. Ferus, J. Cihelka, S. Civis, Formaldehyde in the environment—determination of formaldehyde by laser and photoacoustic detection. Chem. Listy **102**, 417–426 (2008)

M. Ferus, I. Matulkova, L. Juha, S. Civis, Investigation of laser-plasma chemistry in CO-N-2-H2O mixtures using O-18 labeled water. Chem. Phys. Lett. **472**, 14–18 (2009). https://doi.org/10.1016/j.cplett.2009.02.056

M. Ferus, P. Kubelik, K. Kawaguchi et al., HNC/HCN ratio in acetonitrile, formamide, and BrCN discharge. J. Phys. Chem. A **115**, 1885–1899 (2011). https://doi.org/10.1021/jp1107872

M. Ferus, L. Kavan, M. Zukalová et al., Spontaneous and photoinduced conversion of CO_2 on TiO_2 anatase (001)/(101) surfaces. J. Phys. Chem. C **118**, 26845–26850 (2014a). https://doi.org/10.1021/jp5090668

M. Ferus, D. Nesvorný, J. Šponer et al., High-energy chemistry of formamide: a unified mechanism of nucleobase formation. Proc. Natl. Acad. Sci. USA **112**, 657–662 (2014b). https://doi.org/10.1073/pnas.1412072111

M. Ferus, P. Kubelík, A. Knížek et al., High energy radical chemistry formation of HCN-rich atmospheres on early earth. Sci. Rep. **7**(1), 6275 (2017a). https://doi.org/10.1038/s41598-017-06489-1

M. Ferus, F. Pietrucci, A.M. Saitta et al., Formation of nucleobases in a Miller–Urey reducing atmosphere. Proc. Natl. Acad. Sci. **114**, 4306–4311 (2017b). https://doi.org/10.1073/pnas.1700010114

J.J. Flores, W.A. Bonner, On the asymmetric polymerization of aspartic acid enantiomers by kaolin. J. Mol. Evol. **3**, 49–56 (1974). https://doi.org/10.1007/BF01795975

J.J. Flores, W.A. Bonner, G.A. Massey, Asymmetric photolysis of (RS)-leucine with circularly polarized ultraviolet light. J. Am. Chem. Soc. **99**, 3622–3625 (1977). https://doi.org/10.1021/ja00453a018

S. Fonti, G.A. Marzo, Mapping the methane on Mars. Astron. Astrophys. **512**, 6 (2010). https://doi.org/10.1051/0004-6361/200913178

V. Formisano, S. Atreya, T. Encrenaz et al., Detection of methane in the atmosphere of Mars. Science (80-) **306**, 1758–1761 (2004). https://doi.org/10.1126/science.1101732

S.N. Frank, A.J. Bard, Heterogenous photocatalytic oxidation of cyanide ion in aqueous-solutions at TiO_2 powder. J. Am. Chem. Soc. **99**, 303–304 (1977). https://doi.org/10.1021/ja00443a081

D.G. Fraser, D. Fitz, T. Jakschitz, B.M. Rode, Selective adsorption and chiral amplification of amino acids in vermiculite clay-implications for the origin of biochirality. Phys. Chem. Chem. Phys. **13**, 831–838 (2011). https://doi.org/10.1039/C0CP01388A

C. Freissinet, D.P. Glavin, P.R. Mahaffy et al., Organic molecules in the Sheepbed Mudstone, Gale Crater, Mars. J. Geophys. Res. **120**, 495–514 (2015). https://doi.org/10.1002/2014JE004737

M. Fries, A. Christou, D. Archer et al., A cometary origin for martian atmospheric methane. Geochem. Perspect Lett. **2**, 10–22 (2016). https://doi.org/10.7185/geochemlet.1602

A. Fujishima, K. Honda, Electrochemical photolysis of water at a semiconductor electrode. Nature **238**, 37–38 (1972). https://doi.org/10.1038/238037a0

A. Fujishima, K. Hashimoto, T. Watanabe, TiO_2 Photocatalysis: Fundamentals and Applications (BKC, 1999)

M. Gattrell, N. Gupta, A. Co, A review of the aqueous electrochemical reduction of CO_2 to hydrocarbons at copper. J. Electroanal. Chem. **594**, 1–19 (2006). https://doi.org/10.1016/J.JELECHEM.2006.05.013

D.J. Geist, Volcanoes, in *Encyclopedia of Earth System Science*, ed. by W.A. Nierenberg (Academic Press, SanDiego, 1992), pp. 427–436

A. Geminale, V. Formisano, G. Sindoni, Mapping methane in Martian atmosphere with PFS-MEX data. Planet. Space Sci. **59**, 137–148 (2011). https://doi.org/10.1016/j.pss.2010.07.011

D.P. Glavin, C. Freissinet, K.E. Miller et al., Evidence for perchlorates and the origin of chlorinated hydrocarbons detected by SAM at the Rocknest aeolian deposit in Gale Crater. J. Geophys. Res. **118**, 1955–1973 (2013). https://doi.org/10.1002/jgre.20144

P.R. Gordon, M.A. Sephton, Organic matter detection on Mars by pyrolysis-FTIR: an analysis of sensitivity and mineral matrix effects. Astrobiology **16**, 831–845 (2016). https://doi.org/10.1089/ast.2016.1485

S.N. Habisreutinger, L. Schmidt-Mende, J.K. Stolarczyk, Photocatalytic reduction of CO_2 on TiO_2 and other semiconductors. Angew. Chemie-Int. Ed. **52**, 7372–7408 (2013). https://doi.org/10.1002/anie.201207199

K. Hadjiivanov, J. Lamotte, J.C. Lavalley, FTIR study of low-temperature CO adsorption on pure and ammonia-precovered TiO_2 (anatase). Langmuir **13**, 3374–3381 (1997). https://doi.org/10.1021/la962104m

P. Hartogh, C. Jarchow, E. Lellouch et al., Herschel/HIFI observations of Mars: first detection of O_2 at submillimetre wavelengths and upper limits on HCl and H_2O_2. Astron. Astrophys. **521**, L49 (2010). https://doi.org/10.1051/0004-6361/201015160

Z. Hausfather, in *Carbon Br*, Analysis: global CO_2 emissions set to rise 2% in 2017 after three-year "plateau" (2017). https://www.carbonbrief.org/analysis-global-co2-emissions-set-to-rise-2-percent-in-2017-following-three-year-plateau. Accessed 11 Apr 2019

Y. He, O. Dulub, H. Cheng et al., Evidence for the predominance of subsurface defects on reduced anatase $TiO_2(101)$. Phys. Rev. Lett. **102**(10), 106105 (2009). https://doi.org/10.1103/physrevlett.102.106105

A.G. Hearn, The absorption of ozone in the ultra-violet and visible regions of the spectrum. Proc. Phys. Soc. **78**, 932–940 (1961). https://doi.org/10.1088/0370-1328/78/5/340

W. Hebenstreit, N. Ruzycki, G.S. Herman et al., Scanning tunneling microscopy investigation of the TiO_2 anatase (101) surface. Phys. Rev. B **62**, R16334–R16336 (2000)

M.H. Hecht, S.P. Kounaves, R.C. Quinn et al., Detection of perchlorate and the soluble chemistry of martian soil at the phoenix lander site. Science (80) **325**, 64–67 (2009). https://doi.org/10.1126/science.1172466

M.A. Henderson, Formic-acid decomposition on the (110)-microfaceted surface of $TiO_2(100)$—instights derived from 18O-labeling studies. J. Phys. Chem. **99**, 15253–15261 (1995). https://doi.org/10.1021/j100041a048

M.A. Henderson, Structural sensitivity in the dissociation of water on TiO_2 single-crystal surfaces. Langmuir **12**, 5093–5098 (1996). https://doi.org/10.1021/la960360t

M.A. Henderson, W.S. Epling, C.H.F. Peden, C.L. Perkins, Insights into photoexcited electron scavenging processes on TiO2 obtained from studies of the reaction of O-2 with OH groups adsorbed at electronic defects on $TiO_2(110)$. J. Phys. Chem. B **107**, 534–545 (2003). https://doi.org/10.1021/jp0262113

M.R. Hoffmann, S.T. Martin, W.Y. Choi, D.W. Bahnemann, Environmental applications of semiconductor photocatalysis. Chem. Rev. **95**, 69–96 (1995). https://doi.org/10.1021/cr00033a004

H.J. Hofmann, K. Grey, A.H. Hickman, R.I. Thorpe, Origin of 3.45 Ga coniform stromatolites in Warrawoona Group, Western Australia. Geol. Soc. Am. Bull. **111**, 1256–1262 (1999). https://doi.org/10.1130/0016-7606(1999)111%3c1256:OOGCSI%3e2.3.CO;2

S.M. Hörst, R.V. Yelle, A. Buch et al., Formation of amino acids and nucleotide bases in a titan atmosphere simulation experiment. Astrobiology **12**, 809–817 (2012). https://doi.org/10.1089/ast.2011.0623

R. Hu, S. Seager, W. Bains, Photochemistry in terrestrial exoplanet atmospheres.I. Photochemistry model and benchmark cases. Astrophys. J. **761**, 166–195 (2012). https://doi.org/10.1088/0004-637X/761/2/166

R. Hu, D.M. Kass, B.L. Ehlmann, Y.L. Yung, Tracing the fate of carbon and the atmospheric evolution of Mars. Nat. Commun. **6**, 10003 (2015). https://doi.org/10.1038/ncomms10003

R. Hu, A.A. Bloom, P. Gao et al., Hypotheses for near-surface exchange of methane on Mars. Astrobiology **16**, 539–550 (2016). https://doi.org/10.1089/ast.2015.1410

X. Huang, J. Wang, T. Li et al., Review on optofluidic microreactors for artificial photosynthesis. Beilstein. J. Nanotechnol. **9**, 30–41 (2018). https://doi.org/10.3762/bjnano.9.5

C. Huber, G. Wächtershäuser, Activated acetic acid by carbon fixation on (Fe, Ni)S under primordial conditions. Science (80-) **276**, 245–247 (1997). https://doi.org/10.1126/science.276.5310.245

H. Imanaka, B.N. Khare, J.E. Elsila et al., Laboratory experiments of Titan tholin formed in cold plasma at various pressures: implications for nitrogen-containing polycyclic aromatic compounds in Titan haze. Icarus **168**, 344–366 (2004). https://doi.org/10.1016/J.ICARUS.2003.12.014

T. Inoue, A. Fujishima, S. Konishi, K. Honda, Photoelectrocatalytic reduction of carbon-dioxide in aqueous suspensions of semiconductor powders. Nature **277**, 637–638 (1979). https://doi.org/10.1038/277637a0

B.M. Jakosky, R.P. Lin, J.M. Grebowsky et al., The Mars atmosphere and volatile evolution (MAVEN) mission. Space Sci. Rev. **195**, 3–48 (2015). https://doi.org/10.1007/s11214-015-0139-x

T.A.E. Jakschitz, B.M. Rode, Chemical evolution from simple inorganic compounds to chiral peptides. Chem. Soc. Rev. **41**, 5484 (2012). https://doi.org/10.1039/c2cs35073d

C. Jimenez, J. Perriere, C. Palacio et al., Transformation of titanium nitride in oxygen plasma. Thin Solid Films **228**, 247–251 (1993). https://doi.org/10.1016/0040-6090(93)90609-S

A. Johnson, H.J. Cleaves, J.L. Bada, A. Lazcano, The diversity of the original prebiotic soup: re-analyzing the original miller-urey spark discharge experiments. Orig. Life Evol. Biosph. **39**, 240–241 (2009)

C.M. Kalamaras, P. Panagiotopoulou, D.I. Kondarides, A.M. Efstathiou, Kinetic and mechanistic studies of the water-gas shift reaction on Pt/TiO$_2$ catalyst. J. Catal. **264**, 117–129 (2009). https://doi.org/10.1016/j.jcat.2009.03.002

L. Kavan, M. Grätzel, J. Rathouský, A. Zukalb, Nanocrystalline TiO[sub 2] (Anatase) electrodes: surface morphology, adsorption, and electrochemical properties. J. Electrochem. Soc. **143**, 394 (1996). https://doi.org/10.1149/1.1836455

L. Kavan, M. Zukalova, M. Ferus et al., Oxygen-isotope labeled titania: (TiO$_2$)-O-18. Phys. Chem. Chem. Phys. **13**, 11583–11586 (2011). https://doi.org/10.1039/c1cp20775j

A. Knížek, K. Dryahina, P. Španěl et al., Comparative SIFT-MS, GC–MS and FTIR analysis of methane fuel produced in biogas stations and in artificial photosynthesis over acidic anatase TiO$_2$ and montmorillonite. J. Mol. Spectrosc. **648**, 152–160 (2017). https://doi.org/10.1016/j.jms.2017.10.002

V.A. Krasnopolsky, Long-term spectroscopic observations of Mars using IRTF/CSHELL: Mapping of O2 dayglow, CO, and search for CH4. Icarus **190**, 93–102 (2007). https://doi.org/10.1016/J.ICARUS.2007.02.014

V.A. Krasnopolsky, Solar activity variations of thermospheric temperatures on Mars and a problem of CO in the lower atmosphere. Icarus **207**, 638–647 (2010). https://doi.org/10.1016/J.ICARUS.2009.12.036

V.A. Krasnopolsky, J.P. Maillard, T.C. Owen, Detection of methane in the martian atmosphere: evidence for life? Icarus **172**, 537–547 (2004). https://doi.org/10.1016/j.icarus.2004.07.004

E. Krause, M. Bienert, P. Schmieder, H. Wenschuh, The helix-destabilizing propensity scale of d-amino acids: the influence of side chain steric effects. J. Am. Chem. Soc. **122**(20), 4865–4870 (2000). https://doi.org/10.1021/ja9940524

K. Kvenvolden, J. Lawless, K. Pering et al., Evidence for extraterrestrial amino-acids and hydrocarbons in the murchison meteorite. Nature **228**, 923–926 (1970). https://doi.org/10.1038/228923a0

N. Lahav, D. White, S. Chang, Peptide formation in the prebiotic era: thermal condensation of glycine in fluctuating clay environments. Science **201**, 67–69 (1978)

T.D. Lee, C.N. Yang, Question of parity conservation in weak interactions. Phys. Rev. **104**, 254–258 (1956). https://doi.org/10.1103/PhysRev.104.254

J. Lee, D.C. Sorescu, X. Deng, Electron-induced dissociation of CO_2 on TiO_2(110). J. Am. Chem. Soc. **133**, 10066–10069 (2011). https://doi.org/10.1021/ja204077e

N.S. Lewis, D.G. Nocera, Powering the planet: chemical challenges in solar energy utilization. Proc. Natl. Acad. Sci. USA **103**, 15729–15735 (2006). https://doi.org/10.1073/pnas.0603395103

K. Li, X. An, K.H. Park et al., A critical review of CO_2 photoconversion: catalysts and reactors. Catal. Today **224**, 3–12 (2014). https://doi.org/10.1016/j.cattod.2013.12.006

K. Li, B. Peng, T. Peng, Recent advances in heterogeneous photocatalytic CO_2 conversion to solar Fuels. ACS Catal. **6**, 7485–7527 (2016). https://doi.org/10.1021/acscatal.6b02089

X. Li, J. Yu, M. Jaroniec, X. Chen, Cocatalysts for selective photoreduction of CO_2 into solar fuels. Chem. Rev. **119**, 3962–4179 (2019). https://doi.org/10.1021/acs.chemrev.8b00400

L.F. Liao, C.F. Lien, D.L. Shieh et al., FTIR study of adsorption and photoassisted oxygen isotopic exchange of carbon monoxide, carbon dioxide, carbonate, and formate on TiO_2. J. Phys. Chem. B **106**, 11240–11245 (2002). https://doi.org/10.1021/jp0211988

E. Lichtfouse, J. Schwarzbauer, D. Robert (eds.), *Hydrogen Production and Remediation of Carbon and Pollutants* (Springer International Publishing, Cham, 2015)

S.R. Lingampalli, M.M. Ayyub, C.N.R. Rao, Recent progress in the photocatalytic reduction of carbon dioxide. ACS Omega **2**, 2740–2748 (2017). https://doi.org/10.1021/acsomega.7b00721

E.R. Lippincott, R.V. Eck, M.O. Dayhoff, C. Sagan, Thermodynamic equilibria in planetary atmospheres. Astrophys. J. **147**, 753–764 (1967). https://doi.org/10.1086/149051

J.E. Lovelock, A physical basis for life detection experiments. Nature **207**, 568–570 (1965). https://doi.org/10.1038/207568a0

Y. Ma, X. Wang, Y. Jia et al., Titanium dioxide-based nanomaterials for photocatalytic fuel generations. Chem. Rev. **114**, 9987–10043 (2014). https://doi.org/10.1021/cr500008u

E.J. Maginn, What to do with CO_2. J. Phys. Chem. Lett. **1**, 3478–3479 (2010). https://doi.org/10.1021/jz101582c

N. Marom, M. Kim, J.R. Chelikowsky, Structure selection based on high vertical electron affinity for TiO_2 clusters. Phys. Rev. Lett. **108**, 106801 (2012). https://doi.org/10.1103/PhysRevLett.108.106801

Z. Martins, Organic chemistry of carbonaceous meteorites. Elements **7**, 35–40 (2011). https://doi.org/10.2113/gselements.7.1.35

Z. Martins, O. Botta, M.L. Fogel et al., Extraterrestrial nucleobases in the Murchison meteorite. Earth Planet. Sci. Lett. **270**, 130–136 (2008). https://doi.org/10.1016/j.epsl.2008.03.026

Z. Martins, P. Modica, B. Zanda, L.L.S. D'Hendecourt, The amino acid and hydrocarbon contents of the Paris meteorite: Insights into the most primitive CM chondrite. Meteorit. Planet. Sci. **50**, 926–943 (2015). https://doi.org/10.1111/maps.12442

G. Martra, C. Deiana, Y. Sakhno et al., The formation and self-assembly of long prebiotic oligomers produced by the condensation of unactivated amino acids on oxide surfaces. Angew. Chemie. Int. Ed. **53**, 4671–4674 (2014). https://doi.org/10.1002/anie.201311089

H. Martucci, Characterization of perchlorate photostability under simulated martian conditions, in *Proceedings of the National Conference on Undergraduate Research (NCUR) 2012* (Weber State University, Ogden, Utah, 2012), pp. 1359–1363

M. Matsuoka, M. Kitano, M. Takeuchi et al., Photocatalysis for new energy production: Recent advances in photocatalytic water splitting reactions for hydrogen production. Catal. Today **122**, 51–61 (2007). https://doi.org/10.1016/J.CATTOD.2007.01.042

A.F. McKinlay, B.L. Diffey, A reference action spectrum for ultraviolet induced erythema in human skin. CIE J. **6**, 17–22 (1987)

R.V. Mikhaylov, A.A. Lisachenko, V.V. Titov, Investigation of photostimulated oxygen isotope exchange on TiO_2 degussa P25 surface upon UV-Vis irradiation. J. Phys. Chem. C **116**, 23332–23341 (2012). https://doi.org/10.1021/jp305652p

M. Mikkelsen, M. Jørgensen, F.C. Krebs, The teraton challenge. A review of fixation and transformation of carbon dioxide. Energy Environ. Sci. **3**, 43–81 (2010). https://doi.org/10.1039/B912904A

S.L. Miller, A production of amino acids under possible primitive earth conditions. Science (80-) **117**, 528–529 (1953). https://doi.org/10.1126/science.117.3046.528

G.A. Mills, H.C. Urey, the kinetics of isotopic exchange between carbon dioxide, bicarbonate ion, carbonate ion and water1. J. Am. Chem. Soc. **62**(5), 1019–1026 (1940). https://doi.org/10.1021/ja01862a010

J.E. Moores, R.V. Gough, G.M. Martinez et al., Methane seasonal cycle at Gale Crater on Mars consistent with regolith adsorption and diffusion. Nat. Geosci. 1 (2019). https://doi.org/10.1038/s41561-019-0313-y

D.S. Muggli, J.L. Falconer, UV-enhanced exchange of O-2 with H_2O adsorbed on TiO_2. J. Catal. **181**, 155–159 (1999). https://doi.org/10.1006/jcat.1998.2292

M.J. Mumma, G.L. Villanueva, R.E. Novak et al., Strong release of methane on mars in northern summer 2003. Science (80-) **323**, 1041–1045 (2009). https://doi.org/10.1126/science.1165243

H. Nair, M. Allen, A.D. Anbar et al., A photochemical model of the martian atmosphere. Icarus **111**, 124–150 (1994). https://doi.org/10.1006/icar.1994.1137

R. Navarro-González, F.A. Rainey, P. Molina et al., Mars-like soils in the atacama desert, chile, and the dry limit of microbial life. Science (80-) **302**, 1018–1021 (2003)

R. Navarro-Gonzalez, E. Vargas, J. de la Rosa et al., Reanalysis of the viking results suggests perchlorate and organics at midlatitudes on Mars. J. Geophys. Res. **115**, E12010 (2010). https://doi.org/10.1029/2010JE003599

C.D. Neish, A. Somogyi, M.A. Smith, Titan's primordial soup: formation of amino acids via low-temperature hydrolysis of tholins. Astrobiology **10**(3), 337–347 (2010)

V.-H. Nguyen, J.C.S. Wu, Recent developments in the design of photoreactors for solar energy conversion from water splitting and CO_2 reduction. Appl. Catal. A Gen. **550**, 122–141 (2018). https://doi.org/10.1016/J.APCATA.2017.11.002

T. Noël, in *Photochemical Processes in Continuous-Flow Reactors*. World Scientific (Europe) (2017)

M. North, Across the board: michael north on carbon dioxide biorefinery. ChemSusChem **12**(8), 1763–1765 (2019). https://doi.org/10.1002/cssc.201900676

A.P. Nutman, V.R. McGregor, C.R.L. Friend et al., The Itsaq Gneiss Complex of southern West Greenland; the world's most extensive record of early crustal evolution (3900–3600 Ma). Precambrian Res. **78**, 1–39 (1996). https://doi.org/10.1016/0301-9268(95)00066-6

O. Ola, M.M. Maroto-Valer, Review of material design and reactor engineering on TiO_2 photocatalysis for CO_2 reduction. J. Photochem. Photobiol. C Photochem. Rev. **24**, 16–42 (2015). https://doi.org/10.1016/J.JPHOTOCHEMREV.2015.06.001

C. Oze, M. Sharma, Have olivine, will gas: Serpentinization and the abiogenic production of methane on Mars. Geophys. Res. Lett. **32**, L10203 (2005). https://doi.org/10.1029/2005GL022691

S. Pantaleone, P. Ugliengo, M. Sodupe, A. Rimola, When the surface matters: prebiotic peptide-bond formation on the TiO_2 (101) anatase surface through periodic DFT-D2 simulations. Chem Eur. J. **24**, 16292–16301 (2018). https://doi.org/10.1002/chem.201803263

E.T. Parker, H.J. Cleaves, M.P. Callahan et al., Enhanced synthesis of alkyl amino acids in Miller's 1958 H_2S experiment. Orig. Life Evol. Biosph. **41**, 569–574 (2011a). https://doi.org/10.1007/s11084-011-9253-2

E.T. Parker, H.J. Cleaves, J.P. Dworkin et al., Primordial synthesis of amines and amino acids in a 1958 Miller H_2S-rich spark discharge experiment. Proc. Natl. Acad. Sci. USA **108**, 5526–5531 (2011b). https://doi.org/10.1073/pnas.1019191108

P.N. Pearson, M.R. Palmer, Atmospheric carbon dioxide concentrations over the past 60 million years. Nature **406**, 695–699 (2000). https://doi.org/10.1038/35021000

D. Pei, J. Luan, Development of visible light-responsive sensitized photocatalysts. Int. J. Photoenergy **2012**, 1–13 (2012). https://doi.org/10.1155/2012/262831

P. Pichat, H. Courbon, R. Enriquez et al., Light-induced isotopic exchange between O-2 and semiconductor oxides, a characterization method that deserves not to be overlooked. Res. Chem. Intermed. **33**, 239–250 (2007). https://doi.org/10.1163/156856707779238667

S. Pizzarello, R.V. Krishnamurthy, S. Epstein, J.R. Cronin, Isotopic analyses of amino acids from the Murchison meteorite. Geochim. Cosmochim. Acta **55**, 905–910 (1991). https://doi.org/10.1016/0016-7037(91)90350-E

K. Plankensteiner, A. Righi, B.M. Rode, Glycine and diglycine as possible catalytic factors in the prebiotic evolution of peptides. Orig. Life Evol. Biosph. **32**, 225–236 (2002). https://doi.org/10.1023/A:1016523207700

K. Plankensteiner, H. Reiner, B.M. Rode, From earth's primitive atmosphere to chiral peptides—the origin of precursors for life. Chem. Biodivers. **1**, 1308–1315 (2004a). https://doi.org/10.1002/cbdv.200490093

K. Plankensteiner, A. Righi, B.M. Rode et al., Indications towards a stereoselectivity of the salt-induced peptide formation reaction. Inorganica Chim. Acta **357**, 649–656 (2004b). https://doi.org/10.1016/j.ica.2003.06.012

M.W. Powner, B. Gerland, J.D. Sutherland, Synthesis of activated pyrimidine ribonucleotides in prebiotically plausible conditions. Nature **459**, 239–242 (2009). https://doi.org/10.1038/nature08013

S. Protti, A. Albini, N. Serpone, Photocatalytic generation of solar fuels from the reduction of H_2O and CO_2: a look at the patent literature. Phys. Chem. Chem. Phys. **16**, 19790 (2014). https://doi.org/10.1039/C4CP02828G

Z. Qu, G.-J. Kroes, Theoretical study of stable, defect-free $(TiO_2)n$ nanoparticles with n = 10−16. J. Phys. Chem. C **111**(45), 6808–16817 (2007). https://doi.org/10.1021/JP073988T

C.N.R. Rao, S.R. Lingampalli, Generation of hydrogen by visible light-induced water splitting with the use of semiconductors and dyes. Small **12**, 16–23 (2016). https://doi.org/10.1002/smll.201500420

B.M. Rode, Peptide and the origin of life. Peptides **20**, 773–786 (1999). https://doi.org/10.1016/S0196-9781(99)00062-5

G. Ronto, A. Berces, H. Lammer et al., Solar UV irradiation conditions on the surface of Mars. Photochem. Photobiol. **77**, 34–40 (2003). https://doi.org/10.1562/0031-8655(2003)077%3c0034:SUICOT%3e2.0.CO;2

L. Rotelli, J.M. Trigo-Rodríguez, C.E. Moyano-Cambero et al., The key role of meteorites in the formation of relevant prebiotic molecules in a formamide/water environment. Sci. Rep. **6**, 38888 (2016). https://doi.org/10.1038/srep38888

L.S. Rothman, I.E. Gordon, A. Barbe et al., The HITRAN 2008 molecular spectroscopic database. J. Quant. Spectrosc. Radiat. Transf. **110**, 533–572 (2009). https://doi.org/10.1016/j.jqsrt.2009.02.013

C. Sagan, B.N. Khare, Long-wavelength ultraviolet photoproduction of amino acids on the primitive earth. Science (80-) **173**, 417–420 (1971). https://doi.org/10.1126/science.173.3995.417

C. Sagan, W.R. Thompson, R. Carlsom et al., A search for life on Earth from the Galileo spacecraft. Nature **365**, 715–721 (1993). https://doi.org/10.1038/365715a0

S.A. Sandford, F. Salama, L.J. Allamandola et al., Laboratory studies of the newly discovered infrared band at 4705.2 cm-1 (2.1253 mu-m) in the spectrum of Io—the tentative identification of CO_2. Icarus **91**, 125–144 (1991). https://doi.org/10.1016/0019-1035(91)90132-D

S. Sato, Hydrogen and oxygen isotope exchange-reactions over illuminated and nonilluminated TiO_2. J. Phys. Chem. **91**, 2895–2897 (1987). https://doi.org/10.1021/j100295a047

P. Scheiber, M. Fidler, O. Dulub et al., (Sub)Surface mobility of oxygen vacancies at the TiO2 anatase (101) surface. Phys. Rev. Lett. **109**, 136103 (2012). https://doi.org/10.1103/PhysRevLett.109.136103

A.C. Schuerger, R.L. Mancinelli, R.G. Kern et al., Survival of endospores of Bacillus subtilis on spacecraft surfaces under simulated martian environments: implications for the forward contamination of Mars. Icarus **165**, 253–276 (2003). https://doi.org/10.1016/S0019-1035(03) 00200-8

J.D. Schuttlefield, J.B. Sambur, M. Gelwicks et al., Photooxidation of chloride by oxide minerals: implications for perchlorate on mars. J. Am. Chem. Soc. **133**, 17521–17523 (2011). https://doi. org/10.1021/ja2064878

M.G. Schwendinger, B.M. Rode, Possible role of copper and sodium chloride in prebiotic evolution of peptides. Anal. Sci. **5**, 411–414 (1989). https://doi.org/10.2116/analsci.5.411

M.G. Schwendinger, B.M. Rode, Investigations on the mechanism of the salt-induced peptide formation. Orig. Life Evol. Biosph. **22**, 349–359 (1992). https://doi.org/10.1007/BF01809371

S. Seager, W. Bains, J.J. Petkowski, Toward a list of molecules as potential biosignature gases for the search for life on exoplanets and applications to terrestrial biochemistry. Astrobiology **16**, 465–485 (2016). https://doi.org/10.1089/ast.2015.1404

Y.C. Sharma, B. Singh, S.N. Upadhyay, Advancements in development and characterization of biodiesel: a review. Fuel **87**, 2355–2373 (2008). https://doi.org/10.1016/J.FUEL.2008.01.014

A.M. Shaw, *Astrochemistry: From Astronomy to Astrobiology* (Wiley, London, 2006)

N. Shehzad, M. Tahir, K. Johari et al., A critical review on TiO_2 based photocatalytic CO_2 reduction system: strategies to improve efficiency. J. CO_2 Utilization **26**, 98–122 (2018). https://doi.org/10.1016/j.jcou.2018.04.026

I.A. Shkrob, S.D. Chemerisov, T.W. Marin, Photocatalytic decomposition of carboxylated molecules on light-exposed martian regolith and its relation to methane production on Mars. Astrobiology **10**, 425–436 (2010). https://doi.org/10.1089/ast.2009.0433

I.A. Shkrob, N.M. Dimitrijevic, T.W. Marin et al., Heteroatom-transfer coupled photoreduction and carbon dioxide fixation on metal oxides. J. Phys. Chem. C **116**, 9461–9471 (2012a). https://doi.org/10.1021/jp300123z

I.A. Shkrob, T.W. Marin, H. He, P. Zapol, Photoredox reactions and the catalytic cycle for carbon dioxide fixation and methanogenesis on metal oxides. J. Phys. Chem. C **116**, 9450–9460 (2012b). https://doi.org/10.1021/jp300122v

M.L. Smith, M.W. Claire, D.C. Catling, K.J. Zahnle, The formation of sulfate, nitrate and perchlorate salts in the martian atmosphere. Icarus **231**, 51–64 (2014). https://doi.org/10.1016/ j.icarus.2013.11.031

J.P. Smol, Climate change: a planet in flux. Nature **483**, S12–S15 (2012). https://doi.org/10.1038/ 483S12a

Y. Sohn, W. Huang, F. Taghipour, Recent progress and perspectives in the photocatalytic CO_2 reduction of Ti-oxide-based nanomaterials. Appl. Surf. Sci. **396**, 1696–1711 (2017). https:// doi.org/10.1016/J.APSUSC.2016.11.240

D.C. Sorescu, W.A. Al-Saidi, K.D. Jordan, CO_2 adsorption on $TiO_2(101)$ anatase: a dispersion-corrected density functional theory study. J. Chem. Phys. **135**(12), 124701 (2011a). https://doi.org/10.1063/1.3638181

D.C. Sorescu, J. Lee, W.A. Al-Saidi, K.D. Jordan, CO_2 adsorption on $TiO_2(110)$ rutile: insight from dispersion-corrected density functional theory calculations and scanning tunneling microscopy experiments. J. Chem. Phys. **134**(10), 104707 (2011b). https://doi.org/10.1063/1. 3561300

D.C. Sorescu, J. Lee, W.A. Al-Saidi, K.D. Jordan, Coadsorption properties of CO_2 and H_2O on TiO_2 rutile (110): a dispersion-corrected DFT study. J. Chem. Phys. **137**(7), 074704 (2012). https://doi.org/10.1063/1.4739088

D.C. Sorescu, S. Civis, K.D. Jordan, Mechanism of oxygen exchange between CO_2 and $TiO_2(101)$ anatase. J. Phys. Chem. C **118**, 1628–1639 (2014). https://doi.org/10.1021/jp410420e

K. Suriye, P. Praserthdam, B. Jongsomjit, Control of Ti^{3+} surface defect on TiO_2 nanocrystal using various calcination atmospheres as the first step for surface defect creation and its application in photocatalysis. Appl. Surf. Sci. **253**, 3849–3855 (2007). https://doi.org/10.1016/j.apsusc.2006. 08.007

J.D. Sutherland, The origin of life-out of the blue. Angew. Chemie-Int. Ed. **55**, 104–121 (2016)

M. Tahir, B. Tahir, N.A.S. Amin, A. Muhammad, Photocatalytic CO_2 methanation over NiO/In_2O_3 promoted TiO_2 nanocatalysts using H_2O and/or H_2 reductants. Energy Convers. Manag. **119**, 368–378 (2016). https://doi.org/10.1016/J.ENCONMAN.2016.04.057

B. Tahir, M. Tahir, N.A.S. Amin, Tailoring performance of La-modified TiO_2 nanocatalyst for continuous photocatalytic CO_2 reforming of CH_4 to fuels in the presence of H_2O. Energy Convers. Manag. **159**, 284–298 (2018). https://doi.org/10.1016/J.ENCONMAN.2017.12.089

P.J. Thomas, R.D. Hicks, C.F. Chyba, C.P. McKay (eds.), *Comets and the Origin and Evolution of Life* (Springer, Berlin Heidelberg, 2006)

T.L. Thompson, O. Diwald, J.T. Yates, CO_2 as a probe for monitoring the surface defects on TiO2 (110)—temperature-programmed desorption. J. Phys. Chem. B **107**, 11700–11704 (2003). https://doi.org/10.1021/jp030430m

T.L. Thompson, O. Diwald, J.T. Yates, Molecular oxygen-mediated vacancy diffusion on TiO_2(110)—new studies of the proposed mechanism. Chem. Phys. Lett. **393**, 28–30 (2004). https://doi.org/10.1016/j.cplett.2004.05.056

T.T. Veblen, K.R. Young, A.R. Orme, *The physical geopgrahy of South America*, 1st edn. (Oxford University Press, New York, NY, USA, 2007)

J. Wadsworth, C.S. Cockell, Perchlorates on Mars enhance the bacteriocidal effects of UV light. Sci. Rep. **7**, 4662 (2017). https://doi.org/10.1038/s41598-017-04910-3

X. Wang, F. Wang, Y. Sang, H. Liu, Full-spectrum solar-light-activated photocatalysts for light-chemical energy conversion. Adv. Energy Mater. **7**, 1700473 (2017). https://doi.org/10.1002/aenm.201700473

C.R. Webster, P.R. Mahaffy, S.K. Atreya et al., Mars methane detection and variability at Gale crater. Science (80-) **347**, 415–417 (2015). https://doi.org/10.1126/science.1261713

C.R. Webster, P.R. Mahaffy, S.K. Atreya et al., Background levels of methane in Mars' atmosphere show strong seasonal variations. Science (80-) **360**, 1093–1096 (2018). https://doi.org/10.1126/science.aaq0131

C.J. Welch, Formation of highly enantioenriched microenvironments by stochastic sorting of conglomerate crystals: a plausible mechanism for generation of enantioenrichment on the prebiotic earth. Chirality **13**, 425–427 (2001). https://doi.org/10.1002/chir.1055

J.A. Welhan, Origins of methane in hydrothermal systems. Chem. Geol. **71**, 183–198 (1988). https://doi.org/10.1016/0009-2541(88)90114-3

S. Wendt, P.T. Sprunger, E. Lira et al., The Role of Interstitial Sites in the Ti3d Defect State in the Band Gap of Titania. Science (80-) **320**, 1755–1759 (2008). https://doi.org/10.1126/science.1159846

F. Westall, F. Foucher, N. Bost et al., Biosignatures on Mars: what, where, and how? implications for the search for Martian life. Astrobiology **15**, 998–1029 (2015). https://doi.org/10.1089/ast.2015.1374

J.L. White, M.F. Baruch, J.E. Pander et al., Light-driven heterogeneous reduction of carbon dioxide: photocatalysts and photoelectrodes. Chem. Rev. **115**, 12888–12935 (2015). https://doi.org/10.1021/acs.chemrev.5b00370

A.-S. Wong, S.K. Atreya, N.O. Renno, Chemistry related to possible outgassing sources on Mars, in *Sixth International Conference on Mars* (2003a), p. 3

A.S. Wong, S.K. Atreya, T. Encrenaz, Chemical markers of possible hot spots on Mars. J. Geophys. Res. **108**, 5026 (2003b). https://doi.org/10.1029/2002je002003

R. Wordsworth, The climate of early mars. Annu. Rev. Earth Planet. Sci. **44**, 1–31 (2016). https://doi.org/10.1146/annurev-earth-060115-012355

T. Wu, W.E. Kaden, S.L. Anderson, Water on rutile TiO_2(110) and Au/TiO_2(110): effects on an mobility and the isotope exchange reaction. J. Phys. Chem. C **112**, 9006–9015 (2008). https://doi.org/10.1021/jp800521q

X. Yan, L. Tian, X. Chen et al., *Nanomaterials for Photocatalytic Chemistry*, 1st edn. (World Scientific Publishing Co. Pte. Ltd., Singapore, 2017)

Y. Yanagisawa, Oxygen-exchange between CO_2 and metal (Zn and Ti) oxide powders. Energy Convers. Manag. **36**, 443–446 (1995). https://doi.org/10.1016/0196-8904(95)00040-K

Y. Yanagisawa, Y. Ota, Thermal and photo-stimulated desorption of chemisorbed oxygen molecules from titanium-dioxide surfaces. Surf. Sci. **254**, L433–L436 (1991). https://doi.org/10.1016/0039-6028(91)90619-4

Y. Yanagisawa, T. Sumimoto, Oxygen-exchange between CO_2 adsorbate and TiO_2 surfaces. Appl. Phys. Lett. **64**, 3343–3344 (1994). https://doi.org/10.1063/1.111273

X. Yang, F. Gaillard, B. Scaillet, A relatively reduced Hadean continental crust and implications for the early atmosphere and crustal rheology. Earth Planet. Sci. Lett. **393**, 210–219 (2014). https://doi.org/10.1016/j.epsl.2014.02.056

M. Zhang, Q. Wang, C. Chen et al., Oxygen atom transfer in the photocatalytic oxidation of alcohols by TiO_2: oxygen isotope studies. Angew. Chemie-Int. Ed. **48**, 6081–6084 (2009). https://doi.org/10.1002/anie.200900322

R. Zhou, M. Li, S. Wang et al., Low-toxic Mn-doped ZnSe@ZnS quantum dots conjugated with nano-hydroxyapatite for cell imaging. Nanoscale **6**, 14319–14325 (2014). https://doi.org/10.1039/c4nr04473h

Printed in the United States
By Bookmasters